Practice Workbook

ON MY OWN

TEACHER'S EDITION
Grade 1

Harcourt Brace & Company

Orlando • Atlanta • Austin • Boston • San Francisco • Chicago • Dallas • New York • Toronto • London

http://www.hbschool.com

CONTENTS

Name _____

One-to-One Correspondence

Check children's work.

1.

2.

3.

4.

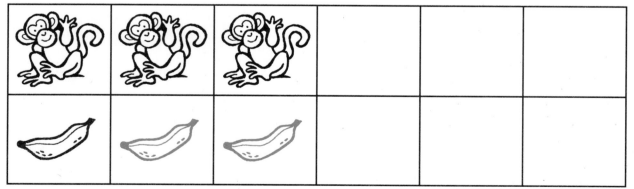

1. Draw one egg for each chicken.
2. Draw one carrot for each rabbit.

3. Draw one ball for each kitten.
4. Draw one banana for each monkey.

Harcourt Brace School Publishers

Name _____

More and Fewer

Answers will vary. Check children's work.

I.—2. Draw flowers to show more flowers than butterflies.

3.—4. Draw leaves to show fewer leaves than bugs.

Numbers Through 5

1.

2.

3.

4.

5.

6.

7.

1. Write the numbers.

2.–7. Count. Write the number that tells how many.

Harcourt Brace School Publishers

Name _____

Numbers Through 9

1.

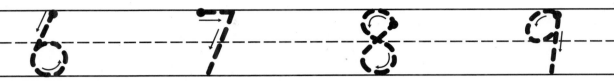

2.

3.

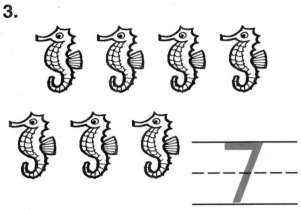

4.

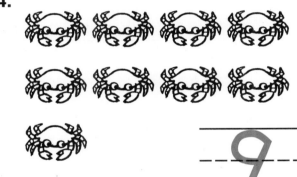

5.

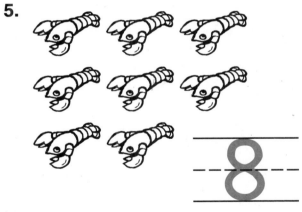

6.

7.

1. Write the numbers.
2.–7. Count. Write the number that tells how many.

Ten

1.

2.

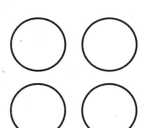

3.

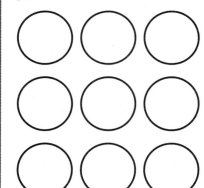

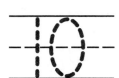

10   8   9

4.

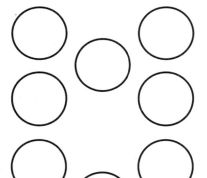

5.

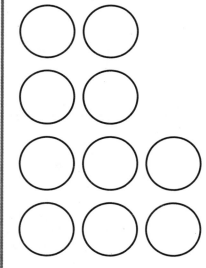

6.

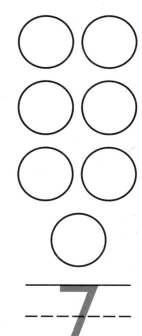

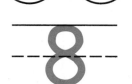

10   10   7

Write the number that tells how many.

Harcourt Brace School Publishers

Greater Than

I.

_ _ 3 _ _

_ _ 6 _ _

2.

7

  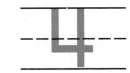

4

3.

8

10

4.

5

 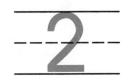

2

Count. Write the numbers.
Compare the groups. Circle the
number that is greater.

Harcourt Brace School Publishers

Name _____

Less Than

I.

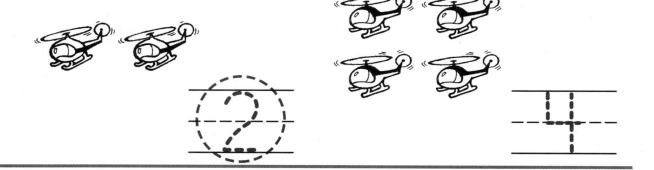

2.

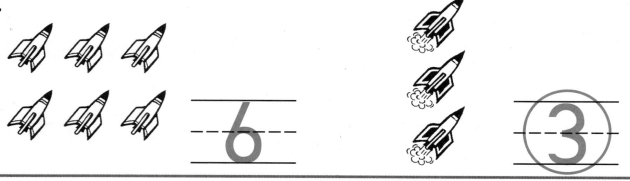

3.

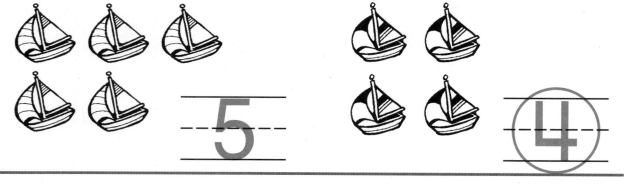

4.

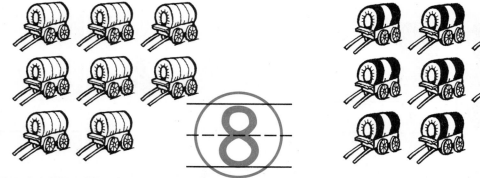

Count. Write the numbers. Compare the groups.
Circle the number that is less.

Harcourt Brace School Publishers

Order Through 10

1.

⭕⭕⭕⭕	⭕⭕⭕⭕⭕	(6 dotted circles)
4	5	6

2.

⭕⭕⭕	⭕⭕⭕⭕	⭕⭕⭕⭕⭕
3	4	5

3.

⭕⭕	⭕⭕⭕	⭕⭕⭕⭕
2	3	4

4.

⭕	⭕⭕	⭕⭕⭕
1	2	3

5.

⭕⭕⭕⭕⭕	⭕⭕⭕⭕⭕⭕	⭕⭕⭕⭕⭕⭕⭕
5	6	7

6.

⭕⭕⭕	⭕⭕⭕⭕	⭕⭕⭕⭕⭕
3	4	5

Draw circles to show the missing number.
Write the number.

Ordinal Numbers   Check children's coloring.

first   second   third   fourth   fifth

1.

yellow

2.

brown

3.

orange

4.

black

5.

red

---

1. Color the third animal yellow.
2. Color the fifth animal brown.
3. Color the second animal orange.

4. Color the fourth animal black.
5. Color the first animal red.

Harcourt Brace School Publishers

Modeling Addition Story Problems

Make up a story.
Use cubes to model the story.
Draw the cubes. Write how many there are in all.

1.

_____   _____   _____ in all

2.

_____   _____   _____ in all

▶ **Problem Solving**

Draw .

Show one way to make 6.
Write how many there are in all.

3.

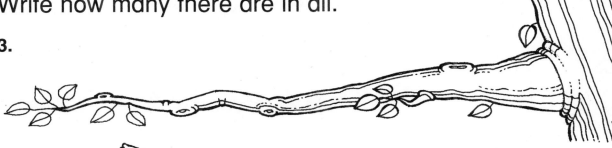

_____   _____   _____ in all

Adding 1

 Vocabulary

Circle the **sum.**

1. 2 + 1 = ③

2. 0 + 1 = ①

Draw and color 1 more. Write the sum.

Check children's drawings.

3.

3 + 1 = __4__
      sum

4.

1 + 1 = __2__

5.

5 + 1 = __6__

6.

2 + 1 = __3__

7.

4 + 1 = __5__

8.

3 + 1 = __4__

Harcourt Brace School Publishers

Adding 2

Draw 2 more balloons.
Color them red.
Write the sum.

Check children's drawings.

1.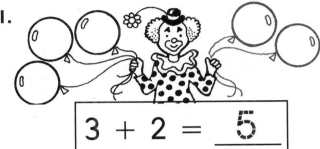

$$3 + 2 = \underline{5}$$

2.

$$1 + 2 = \underline{3}$$

3.

$$2 + 2 = \underline{4}$$

4.

$$4 + 2 = \underline{6}$$

5.

$$3 + 2 = \underline{5}$$

6.

$$2 + 2 = \underline{4}$$

▶ **Problem Solving**

Which group has 2 more than 3? Circle the group.

7.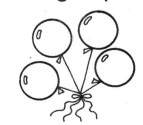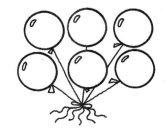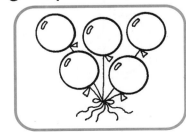

Using Pictures to Add

Write the sum.

1.

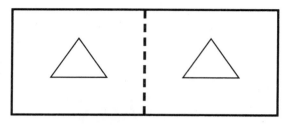

$1 + 1 = \underline{2}$

2.

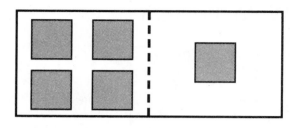

$4 + 1 = \underline{5}$

3.

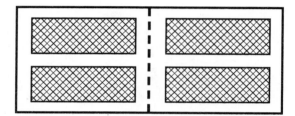

$2 + 2 = \underline{4}$

4.

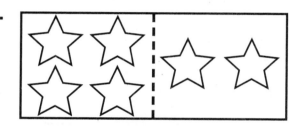

$4 + 2 = \underline{6}$

5.

$3 + 2 = \underline{5}$

6.

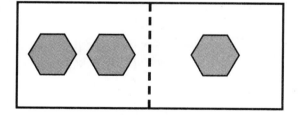

$2 + 1 = \underline{3}$

7.

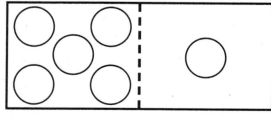

$5 + 1 = \underline{6}$

8.

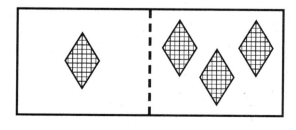

$1 + 3 = \underline{4}$

Harcourt Brace School Publishers

Writing Addition Sentences

Write the addition sentence.

1.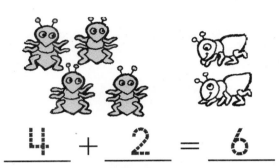

    __4__ + __2__ = __6__

2.

    __2__ + __1__ = __3__

3.

    __3__ + __2__ = __5__

4.

    __1__ + __3__ = __4__

5.

    __2__ + __2__ = __4__

6.

    __1__ + __1__ = __2__

▶ **Problem Solving**

Check children's drawings.

Draw. Write the addition sentence.

7. Draw 1 yellow 🦋.
   Draw 2 blue 🦋.

    __1__ + __2__ = __3__

8. Draw 3 orange 🦋.
   Draw 1 green 🦋.

    __3__ + __1__ = __4__

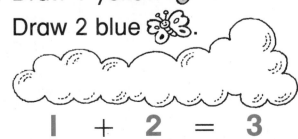 

Modeling Subtraction Story Problems

Tell a story to a friend.   **Check children's drawings.**
Use counters to model the story.
Draw the counters. Write how many are left.

1.

_____ swimmers   _____ go away   _____ are left

2.

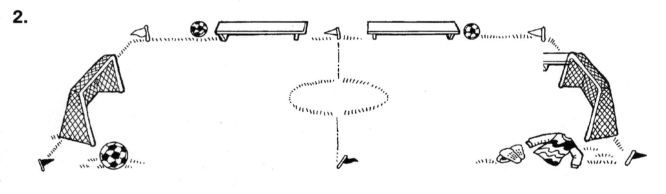

_____ soccer players   _____ go away   _____ are left

3.

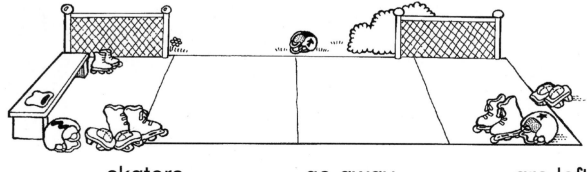

_____ skaters   _____ go away   _____ are left

Harcourt Brace School Publishers

Subtracting 1

▶ **Vocabulary**

Circle the **subtraction sentence**.

1.

$2 + 1 = 3$ $\boxed{2 - 1 = 1}$

Cross out 1 picture. Write how many are left.

2. $3 - 1 = \underline{2}$

3. $4 - 1 = \underline{3}$

4. $2 - 1 = \underline{1}$

5. $5 - 1 = \underline{4}$

▶ **Problem Solving**

Tell a story. Write how many are left.

6.

$5 - 1 = \underline{4}$

7.

$4 - 1 = \underline{3}$

Subtracting 2

Cross out pictures to show the
subtraction sentence. Write how many are left.

1.
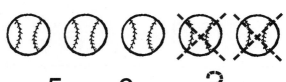

$5 - 2 = \underline{3}$

2.

$2 - 1 = \underline{1}$

3.

$1 - 1 = \underline{0}$

4.

$5 - 2 = \underline{3}$

5.

$4 - 2 = \underline{2}$

6.

$3 - 2 = \underline{1}$

7.
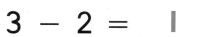

$6 - 1 = \underline{5}$

8.

$1 - 1 = \underline{0}$

9.

$6 - 2 = \underline{4}$

10.

$4 - 2 = \underline{2}$

Harcourt Brace School Publishers

Writing Subtraction Sentences

▶ **Vocabulary**

Circle the **difference**.

1. 4 – 2 = ②

2. 6 – 2 = ④

Write a subtraction sentence to show the difference.

3.

_4_ – _2_ = _2_
difference

4.

_3_ – _1_ = _2_

5.

_6_ – _3_ = _3_

6.

_5_ – _2_ = _3_

7.

_2_ – _1_ = _1_

8.

_4_ – _3_ = _1_

Problem Solving • Make a Model

Add or subtract. Use counters.
Draw the counters.

Check children's drawings.

1. 4 ducks are swimming.
I more comes.
How many in all?

 ___5___ ducks

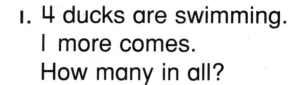

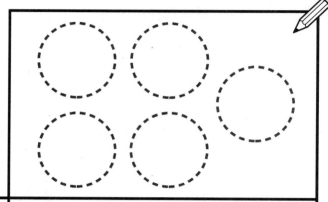

2. 6 kittens are playing.
4 run away.
How many are left?

 ___2___ kittens

3. 3 bees are on a flower.
2 more come.
How many in all?

 ___5___ bees

4. 3 turtles are on a log.
I goes into the water.
How many are left?

 ___2___ turtles

Harcourt Brace School Publishers

Order Property

Use two-color counters and Workmat 2
to find each sum. Circle the two
problems that have the same sum.

1.

$(2 + 1 =)$ __3__ $1 + 0 =$ __1__ $(1 + 2 =)$ __3__

2.

$0 + 3 =$ __3__ $(3 + 1 =)$ __4__ $(1 + 3 =)$ __4__

3.

$(2 + 4 =)$ __6__ $4 + 1 =$ __5__ $(4 + 2 =)$ __6__

4.

$3 + 1 =$ __4__ $(3 + 0 =)$ __3__ $(0 + 3 =)$ __3__

5.

$(3 + 2 =)$ __5__ $(2 + 3 =)$ __5__ $2 + 0 =$ __2__

6.

$(5 + 1 =)$ __6__ $(1 + 5 =)$ __6__ $3 + 4 =$ __7__

▶ **Problem Solving**

Circle the cubes that show the same number.

7.

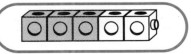  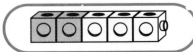

Addition Combinations

Use two-color counters and Workmat 2.
Find ways to make the sums.

Answers may vary. Possible answers:

7+0,
0+7,
6+1,
1+6,
5+2,
2+5,
4+3,
3+4

1. __2__ + __5__ = 7

2. __3__ + __4__ = 7

3. _____ + _____ = 7

4. _____ + _____ = 7

5. _____ + _____ = 7

6. _____ + _____ = 7

Answers may vary. Possible answers:

8+0,
0+8,
7+1,
1+7,
6+2,
2+6,
5+3,
3+5,
4+4

7. _____ + _____ = 8

8. _____ + _____ = 8

9. _____ + _____ = 8

10. _____ + _____ = 8

11. _____ + _____ = 8

12. _____ + _____ = 8

▶ **Problem Solving**

Draw fish in each bowl to make the sum.
Write the numbers.

Possible answers are given.

13.

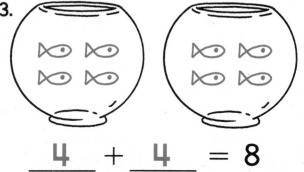

__4__ + __4__ = 8

14.

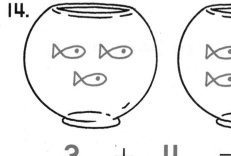

__3__ + __4__ = 7

Harcourt Brace School Publishers

More Addition Combinations

Use two colors of cubes to find
ways to make the sums.
Color to show the ways.

1.
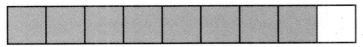

$\underline{8} + \underline{1} = 9$

2.
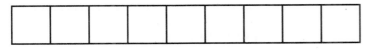

$\underline{} + \underline{} = 9$

3.
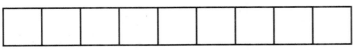

$\underline{} + \underline{} = 9$

4.
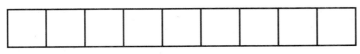

$\underline{} + \underline{} = 9$

5.
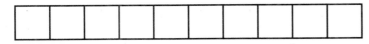

$\underline{9} + \underline{1} = 10$

6.
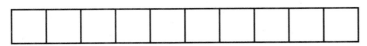

$\underline{} + \underline{} = 10$

7.
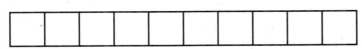

$\underline{} + \underline{} = 10$

8.

$\underline{} + \underline{} = 10$

Horizontal and Vertical Addition

Complete.

1.

$\underline{2} + \underline{1} = \underline{3}$

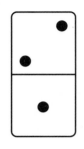

$$\begin{array}{r} 2 \\ +\ 1 \\ \hline 3 \end{array}$$

2.

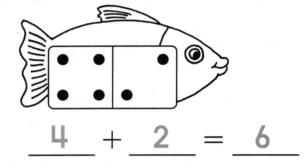

$\underline{4} + \underline{2} = \underline{6}$

$$\begin{array}{r} 4 \\ +\ 2 \\ \hline 6 \end{array}$$

3.

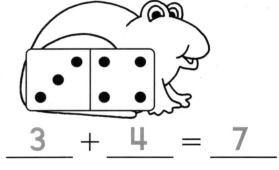

$\underline{3} + \underline{4} = \underline{7}$

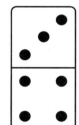

$$\begin{array}{r} 3 \\ +\ 4 \\ \hline 7 \end{array}$$

▶ **Problem Solving**

Write the sum.

4. Circle the sum that is greater.

$8 + 2 = \boxed{10}$

$8 + 1 = \underline{9}$

5. Circle the sum that is less.

$7 + 1 = \underline{8}$

$7 + 0 = \boxed{7}$

Harcourt Brace School Publishers

Problem Solving • Make a Model

► **Vocabulary**

Circle the **cent** sign.

1.   + $ −   （¢）

Mark an **X** on the **penny**.

2.

Use pennies to show each price.
Draw them. Write the total amount.   **Check children's drawings.**

3.

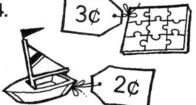 _7_ ¢

4.  _5_ ¢

5.  _9_ ¢

6.  _6_ ¢

Counting On 1 and 2

▶ **Vocabulary**

Count on to add.

1.
$$\begin{array}{r} 4 \\ +\ 1 \\ \hline 5 \end{array}$$ Count on 1. **5** $$\begin{array}{r} 7 \\ +\ 2 \\ \hline 9 \end{array}$$ Count on 2. **8,9** $$\begin{array}{r} 6 \\ +\ 1 \\ \hline 7 \end{array}$$ Count on 1. **7**

2.
$$\begin{array}{r} 4 \\ +\ 2 \\ \hline 6 \end{array}$$ $$\begin{array}{r} 5 \\ +\ 1 \\ \hline 6 \end{array}$$ $$\begin{array}{r} 8 \\ +\ 2 \\ \hline 10 \end{array}$$ $$\begin{array}{r} 6 \\ +\ 1 \\ \hline 7 \end{array}$$ $$\begin{array}{r} 2 \\ +\ 1 \\ \hline 3 \end{array}$$

3.
$$\begin{array}{r} 2 \\ +\ 2 \\ \hline 4 \end{array}$$ $$\begin{array}{r} 1 \\ +\ 2 \\ \hline 3 \end{array}$$ $$\begin{array}{r} 8 \\ +\ 1 \\ \hline 9 \end{array}$$ $$\begin{array}{r} 1 \\ +\ 1 \\ \hline 2 \end{array}$$ $$\begin{array}{r} 7 \\ +\ 2 \\ \hline 9 \end{array}$$

4.
$$\begin{array}{r} 5 \\ +\ 1 \\ \hline 6 \end{array}$$ $$\begin{array}{r} 7 \\ +\ 2 \\ \hline 9 \end{array}$$ $$\begin{array}{r} 4 \\ +\ 1 \\ \hline 5 \end{array}$$ $$\begin{array}{r} 2 \\ +\ 2 \\ \hline 4 \end{array}$$ $$\begin{array}{r} 6 \\ +\ 1 \\ \hline 7 \end{array}$$

▶ **Problem Solving**

Tell a story. Write the addition sentence.

5.

$$\underline{\ 3\ } \ \bigoplus \ \underline{\ 1\ } \ \bigodot \ \underline{\ 4\ }$$

6.

$$\underline{\ 7\ } \ \bigoplus \ \underline{\ 2\ } \ \bigodot \ \underline{\ 9\ }$$

Harcourt Brace School Publishers

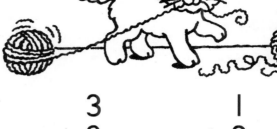

Counting On 3

Count on to add.

1.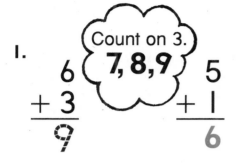

$\begin{array}{r} 6 \\ +3 \\ \hline 9 \end{array}$ Count on 3. **7, 8, 9**

$\begin{array}{r} 5 \\ +1 \\ \hline 6 \end{array}$
$\begin{array}{r} 3 \\ +2 \\ \hline 5 \end{array}$
$\begin{array}{r} 1 \\ +3 \\ \hline 4 \end{array}$
$\begin{array}{r} 5 \\ +2 \\ \hline 7 \end{array}$

2.
$\begin{array}{r} 7 \\ +3 \\ \hline 10 \end{array}$
$\begin{array}{r} 6 \\ +2 \\ \hline 8 \end{array}$
$\begin{array}{r} 3 \\ +3 \\ \hline 6 \end{array}$
$\begin{array}{r} 6 \\ +1 \\ \hline 7 \end{array}$
$\begin{array}{r} 1 \\ +2 \\ \hline 3 \end{array}$

3.
$\begin{array}{r} 4 \\ +3 \\ \hline 7 \end{array}$
$\begin{array}{r} 4 \\ +2 \\ \hline 6 \end{array}$
$\begin{array}{r} 2 \\ +2 \\ \hline 4 \end{array}$
$\begin{array}{r} 2 \\ +3 \\ \hline 5 \end{array}$
$\begin{array}{r} 7 \\ +1 \\ \hline 8 \end{array}$

4.
$\begin{array}{r} 6 \\ +3 \\ \hline 9 \end{array}$
$\begin{array}{r} 5 \\ +3 \\ \hline 8 \end{array}$
$\begin{array}{r} 2 \\ +1 \\ \hline 3 \end{array}$
$\begin{array}{r} 8 \\ +1 \\ \hline 9 \end{array}$
$\begin{array}{r} 7 \\ +2 \\ \hline 9 \end{array}$

▶ **Problem Solving**

Solve.

5. Sara has 5 pencils. Tom has 3 pencils. How many pencils do Sara and Tom have in all?

___8___ pencils

Check children's work.

Name _____

Doubles

▶ **Vocabulary**

Circle the **doubles**.

1. 3 + 2 = 5  3 + 3 = 6

Make each picture show a double.
Write the doubles fact.

Check children's drawings.

2.

__5__ + __5__ = __10__

3.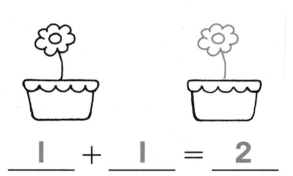

__1__ + __1__ = __2__

4.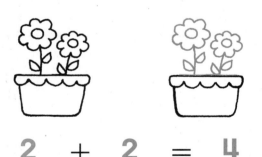

__2__ + __2__ = __4__

5.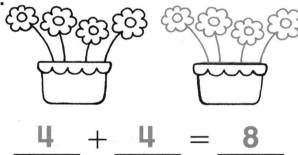

__4__ + __4__ = __8__

▶ **Problem Solving**

6. Jesse has 3 apples.
Matt has double this amount.
How many apples
does Matt have?

__6__ apples

Check children's work.

Addition Facts Practice

Write the sum. Circle the doubles.

1.

$3 + 4 = \underline{7}$ $5 + 1 = \underline{6}$ $(4 + 4 = \underline{8})$

2.

$2 + 4 = \underline{6}$ $(3 + 3 = \underline{6})$ $4 + 1 = \underline{5}$

3.

$0 + 2 = \underline{2}$ $6 + 2 = \underline{8}$ $(1 + 1 = \underline{2})$

4.

$(2 + 2 = \underline{4})$ $3 + 2 = \underline{5}$ $7 + 1 = \underline{8}$

5.

$\begin{array}{r} 3 \\ +1 \\ \hline 4 \end{array}$ $\begin{array}{r} 1 \\ +1 \\ \hline 2 \end{array}$ $\begin{array}{r} 5 \\ +5 \\ \hline 10 \end{array}$ $\begin{array}{r} 6 \\ +2 \\ \hline 8 \end{array}$ $\begin{array}{r} 6 \\ +3 \\ \hline 9 \end{array}$

6.

$\begin{array}{r} 3 \\ +3 \\ \hline 6 \end{array}$ $\begin{array}{r} 4 \\ +1 \\ \hline 5 \end{array}$ $\begin{array}{r} 2 \\ +5 \\ \hline 7 \end{array}$ $\begin{array}{r} 4 \\ +4 \\ \hline 8 \end{array}$ $\begin{array}{r} 2 \\ +1 \\ \hline 3 \end{array}$

▶ **Problem Solving**

7. Doug has 3 cars.
Jim has 3 cars.
How many cars do Doug
and Jim have in all?

Check children's work.

$\underline{6}$ cars

Problem Solving • Draw a Picture

Draw pictures to solve.

Check children's drawings.

I. 4 kittens are white.
2 kittens are black.
How many kittens in all?

_6_

---

2. David has 6 yo-yos.
He gives away 2.
How many are left?

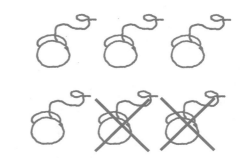

_4_

---

3. Jeremy has 3 red balloons.
Ross has 3 blue balloons.
How many balloons in all?

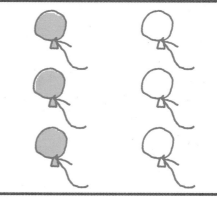

_6_

---

4. Heather has 7 flowers.
She gives away 2.
How many are left?

_5_

Subtraction Combinations

Use counters. Write ways to subtract.

Order of answers may vary.
Possible answers are given.

1. $7 - \underline{7} = \underline{0}$ 2. $8 - \underline{3} = \underline{5}$

3. $7 - \underline{5} = \underline{2}$ 4. $8 - \underline{4} = \underline{4}$

5. $7 - \underline{1} = \underline{6}$ 6. $8 - \underline{7} = \underline{1}$

Subtract.

7.

$$\begin{array}{r} 7 \\ -4 \\ \hline 3 \end{array} \quad \begin{array}{r} 6 \\ -2 \\ \hline 4 \end{array} \quad \begin{array}{r} 7 \\ -7 \\ \hline 0 \end{array} \quad \begin{array}{r} 8 \\ -4 \\ \hline 4 \end{array} \quad \begin{array}{r} 7 \\ -6 \\ \hline 1 \end{array} \quad \begin{array}{r} 8 \\ -3 \\ \hline 5 \end{array}$$

8.

$$\begin{array}{r} 5 \\ -3 \\ \hline 2 \end{array} \quad \begin{array}{r} 8 \\ -7 \\ \hline 1 \end{array} \quad \begin{array}{r} 7 \\ -5 \\ \hline 2 \end{array} \quad \begin{array}{r} 3 \\ -1 \\ \hline 2 \end{array} \quad \begin{array}{r} 7 \\ -0 \\ \hline 7 \end{array} \quad \begin{array}{r} 7 \\ -1 \\ \hline 6 \end{array}$$

9.

$$\begin{array}{r} 6 \\ -3 \\ \hline 3 \end{array} \quad \begin{array}{r} 8 \\ -0 \\ \hline 8 \end{array} \quad \begin{array}{r} 7 \\ -2 \\ \hline 5 \end{array} \quad \begin{array}{r} 8 \\ -8 \\ \hline 0 \end{array} \quad \begin{array}{r} 9 \\ -7 \\ \hline 2 \end{array} \quad \begin{array}{r} 8 \\ -5 \\ \hline 3 \end{array}$$

▶ **Problem Solving**

Which answer will be less than 5?
Circle the problem. Solve to check.

10. $\boxed{8 - 4 = \underline{4}}$ $8 - 2 = \underline{6}$ $8 - 1 = \underline{7}$

More Subtraction Combinations

Use counters. Write ways to subtract. Order of answers may vary.
Possible answers are given.

1. $9 - 0 = 9$

2. $10 - 4 = 6$

3. $9 - 4 = 5$

4. $10 - 1 = 9$

5. $9 - 2 = 7$

6. $10 - 7 = 3$

7. $9 - 1 = 8$

8. $10 - 5 = 5$

9. $9 - 6 = 3$

10. $10 - 8 = 2$

11. $9 - 5 = 4$

12. $10 - 2 = 8$

13. $9 - 7 = 2$

14. $10 - 3 = 7$

15. $9 - 3 = 6$

16. $10 - 9 = 1$

Harcourt Brace School Publishers

Vertical Subtraction

Complete.

1.

4 – 1 = 3

$$\begin{array}{r} 4 \\ -\ 1 \\ \hline 3 \end{array}$$

2.

6 – 2 = 4

$$\begin{array}{r} 6 \\ -\ 2 \\ \hline 4 \end{array}$$

3.

5 – 4 = 1

$$\begin{array}{r} 5 \\ -\ 4 \\ \hline 1 \end{array}$$

▶ **Problem Solving**

Add or subtract. Circle the answer that is greater.

4. Sam has 7 balloons. He breaks 2. How many are left?

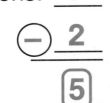

$$\begin{array}{r} 7 \\ \ominus\ 2 \\ \hline \boxed{5} \end{array}$$

Liz has 3 fish. She buys 1 more. How many does she have?

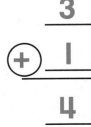

$$\begin{array}{r} 3 \\ \oplus\ 1 \\ \hline 4 \end{array}$$

Name _____

Fact Families

▶ Vocabulary

Circle the sentence that does not belong in the **fact family**.

1.

$6 + 3 = 9$ $\boxed{9 - 5 = 4}$

$3 + 6 = 9$ $9 - 6 = 3$

$9 - 3 = 6$

Add or subtract. Write the numbers in the fact family.

2.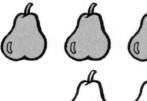

$$\begin{array}{c} 4 \\ + 2 \\ \hline 6 \end{array} \quad \begin{array}{c} 2 \\ + 4 \\ \hline 6 \end{array} \quad \begin{array}{c} 6 \\ - 2 \\ \hline 4 \end{array} \quad \begin{array}{c} 6 \\ - 4 \\ \hline 2 \end{array}$$

$\boxed{4}$ $\boxed{2}$ $\boxed{6}$

3.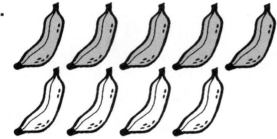

$$\begin{array}{c} 5 \\ + 4 \\ \hline 9 \end{array} \quad \begin{array}{c} 4 \\ + 5 \\ \hline 9 \end{array} \quad \begin{array}{c} 9 \\ - 4 \\ \hline 5 \end{array} \quad \begin{array}{c} 9 \\ - 5 \\ \hline 4 \end{array}$$

$\boxed{5}$ $\boxed{4}$ $\boxed{9}$

4.

$$\begin{array}{c} 5 \\ + 3 \\ \hline 8 \end{array} \quad \begin{array}{c} 3 \\ + 5 \\ \hline 8 \end{array} \quad \begin{array}{c} 8 \\ - 3 \\ \hline 5 \end{array} \quad \begin{array}{c} 8 \\ - 5 \\ \hline 3 \end{array}$$

$\boxed{5}$ $\boxed{3}$ $\boxed{8}$

▶ Problem Solving

5. Tell a story. Write the numbers in the fact family.

$\boxed{7}$ $\boxed{3}$ $\boxed{10}$

ON MY OWN

Subtracting to Compare

Draw lines to match. Subtract to compare.
Write how many more.

1.

$7 - 5 =$ ___2___

___2___ more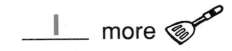

2.

$6 - 2 =$ ___4___

___4___ more

3.

$5 - 4 =$ ___1___

___1___ more

4.

$4 - 1 =$ ___3___

___3___ more

▶ **Problem Solving**

Solve.

5. You have 8 🌼.   ___8___

   You have 6 🌷. $-$ ___6___

   How many more
   🌼 do you have?   ___2___

6. You have 6 🔒.   ___6___

   You have 1 🔑. $-$ ___1___

   How many more
   🔑 do you need?   ___5___

Counting Back 1 and 2

▶ **Vocabulary**

Use the number line.
Count back to subtract.

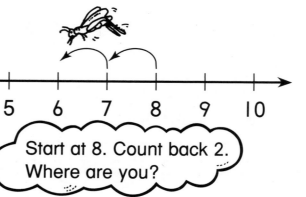

1.
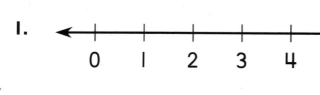

$8 - 2 = \underline{6}$

Start at 8. Count back 2.
Where are you?

2.

$5 - 1 = \underline{4}$

3.

$3 - 2 = \underline{1}$

4.

$4 - 2 = \underline{2}$

5.

$4 - 1 = \underline{3}$

6.

$10 - 2 = \underline{8}$

7.

$9 - 2 = \underline{7}$

▶ **Problem Solving**

Solve.

8. Sophie had 5 balloons.
3 blew away.
How many balloons does
she have left? $\underline{2}$ balloons

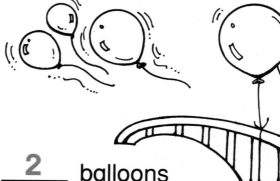

Counting Back 3

Use the number line. Count back to subtract.

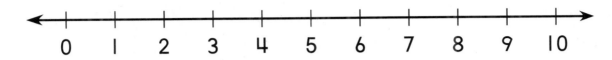

$$0 \quad 1 \quad 2 \quad 3 \quad 4 \quad 5 \quad 6 \quad 7 \quad 8 \quad 9 \quad 10$$

1. $8 - 3 = \underline{5}$　　$9 - 3 = \underline{6}$　　$6 - 3 = \underline{3}$

2. $4 - 2 = \underline{2}$　　$7 - 1 = \underline{6}$　　$9 - 2 = \underline{7}$

3. $3 - 1 = \underline{2}$　　$3 - 3 = \underline{0}$　　$2 - 1 = \underline{1}$

4.
$$\begin{array}{r} 7 \\ -3 \\ \hline 4 \end{array} \qquad \begin{array}{r} 8 \\ -1 \\ \hline 7 \end{array} \qquad \begin{array}{r} 4 \\ -2 \\ \hline 2 \end{array} \qquad \begin{array}{r} 8 \\ -3 \\ \hline 5 \end{array} \qquad \begin{array}{r} 7 \\ -1 \\ \hline 6 \end{array} \qquad \begin{array}{r} 4 \\ -3 \\ \hline 1 \end{array}$$

5.
$$\begin{array}{r} 5 \\ -2 \\ \hline 3 \end{array} \qquad \begin{array}{r} 2 \\ -1 \\ \hline 1 \end{array} \qquad \begin{array}{r} 3 \\ -3 \\ \hline 0 \end{array} \qquad \begin{array}{r} 6 \\ -2 \\ \hline 4 \end{array} \qquad \begin{array}{r} 4 \\ -3 \\ \hline 1 \end{array} \qquad \begin{array}{r} 6 \\ -1 \\ \hline 5 \end{array}$$

▶ **Problem Solving**　　**Possible answers: $10-7=3$; $10-3=7$**

Solve.

6. Use these numbers.
   Write a subtraction sentence.

10　7　3

_____ − _____ = _____

Subtracting Zero

Subtract. Circle all the zero facts.

1.

$$\begin{array}{r} 6 \\ -6 \\ \hline 0 \end{array}$$

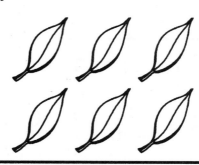

$$\begin{array}{r} 6 \\ -0 \\ \hline 6 \end{array}$$

2.

$$\begin{array}{r} 9 \\ -9 \\ \hline 0 \end{array} \qquad \begin{array}{r} 5 \\ -0 \\ \hline 5 \end{array} \qquad \begin{array}{r} 8 \\ -3 \\ \hline 5 \end{array} \qquad \begin{array}{r} 5 \\ -5 \\ \hline 0 \end{array} \qquad \begin{array}{r} 3 \\ -3 \\ \hline 0 \end{array} \qquad \begin{array}{r} 7 \\ -3 \\ \hline 4 \end{array}$$

3.

$$\begin{array}{r} 7 \\ -2 \\ \hline 5 \end{array} \qquad \begin{array}{r} 8 \\ -8 \\ \hline 0 \end{array} \qquad \begin{array}{r} 5 \\ -4 \\ \hline 1 \end{array} \qquad \begin{array}{r} 4 \\ -4 \\ \hline 0 \end{array} \qquad \begin{array}{r} 5 \\ -1 \\ \hline 4 \end{array} \qquad \begin{array}{r} 3 \\ -0 \\ \hline 3 \end{array}$$

4.

$$\begin{array}{r} 4 \\ -1 \\ \hline 3 \end{array} \qquad \begin{array}{r} 7 \\ -7 \\ \hline 0 \end{array} \qquad \begin{array}{r} 3 \\ -2 \\ \hline 1 \end{array} \qquad \begin{array}{r} 10 \\ -1 \\ \hline 9 \end{array} \qquad \begin{array}{r} 2 \\ -1 \\ \hline 1 \end{array} \qquad \begin{array}{r} 9 \\ -9 \\ \hline 0 \end{array}$$

▶ **Problem Solving**

Write the subtraction sentence.

5.

$$\underline{\ \ 5\ \ } - \underline{\ \ 5\ \ } = \underline{\ \ 0\ \ }$$

Name _____

Facts Practice

Add or subtract. Fill in the numbers.

1.

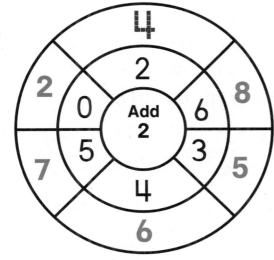

2.

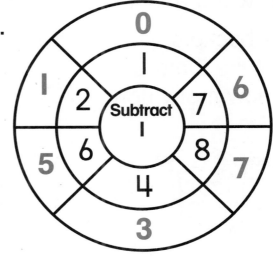

3.

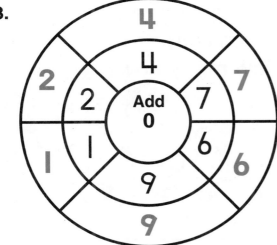

4.

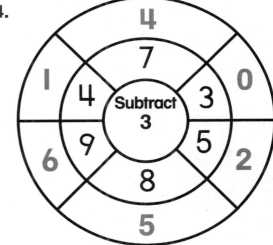

▶ **Problem Solving**

Circle **add** or **subtract**. Solve.

5. Carol has 6 flowers.
She picks 4 more.
How many flowers does she have in all? ___10___ flowers

    (**add**)   **subtract**

Harcourt Brace School Publishers

Problem Solving • Draw a Picture

Add or subtract.   **Check children's drawings.**
Draw more things, or cross things out.

1. 3 buckets are in a row.
   Make 2 more.
   How many now? ___5___

2. 6 starfish are on the beach.
   2 swim away.
   How many now? ___4___

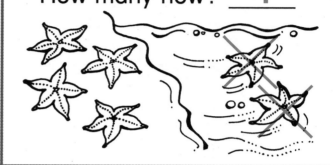

3. 1 chair is in the sand.
   Make 1 more.
   How many now? ___2___

4. 4 seagulls are walking.
   1 flies away.
   How many now? ___3___

5. 5 puppies are sleeping.
   3 run away.
   How many now? ___2___

6. 2 balls are in the yard.
   Make 2 more.
   How many now? ___4___

Harcourt Brace School Publishers

Solid Figures

▶ **Vocabulary**

Draw a line to match each word with a figure.

1.
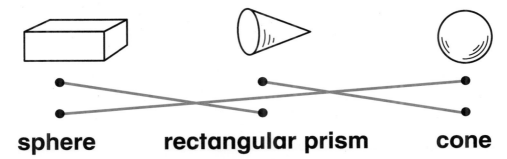

sphere   **rectangular prism**   **cone**

Color the figures that have the same shape.

2.
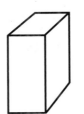

3.

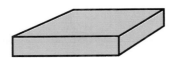

4.

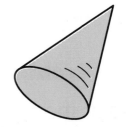

▶ **Problem Solving**

5. I am a solid figure.
   I can roll. Circle me.

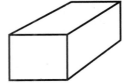

Harcourt Brace School Publishers

More Solid Figures

▶ **Vocabulary**

Draw a line to match each word with a figure.

1.

cylinder

cube

pyramid

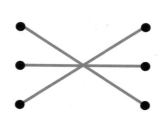

Color the objects that have the same shape.

2.

3.

4.

▶ **Problem Solving**

Color the figures that are cylinders.
Mark an **X** on the figures that are not cylinders.

5.
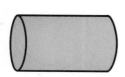

Sorting Solid Figures

 **stack**  **slide**  **roll**

Color each figure that will stack.

1.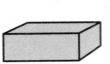

Color each figure that will slide.

2.

Color each figure that will roll.

3.

Color each figure that will roll and stack.

4.

▶ **Problem Solving**

Color the figure that will stack, slide, and roll.

5.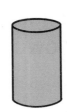

More Sorting Solid Figures

Circle each figure that goes with the sentence.

1. No face is flat.

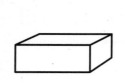 

2. All faces are flat.

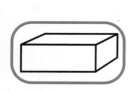 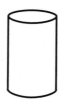    

3. Only 1 face is flat.

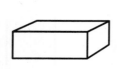   

4. Only 2 faces are flat.

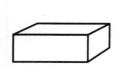 

 Problem Solving

5. Find all the △.
   Write how many.

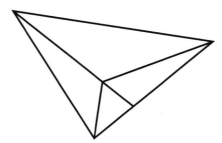

 8

Harcourt Brace School Publishers

Problem Solving • Make a Model

Build the model. Write how many cubes you used.

1.

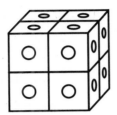

_8_ cubes

2.

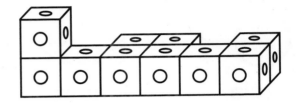

_10_ cubes

3.

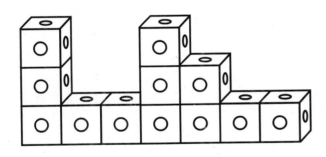

_12_ cubes

4.

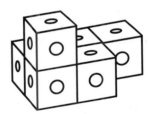

_6_ cubes

5.

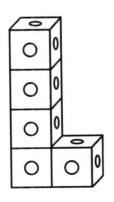

_5_ cubes

6.

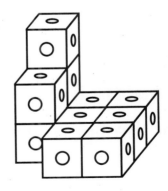

_11_ cubes

Plane Figures

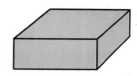 **Vocabulary**

Color each **flat face**.

1.

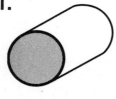 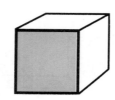 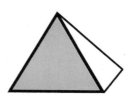 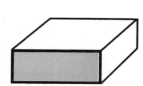

circle   **square**   **triangle**   **rectangle**

2. Color the triangles.

3. Color the squares.

 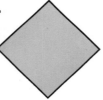

4. Color the circles.

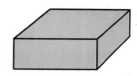 **Problem Solving**

5. Color the figure that has all **flat faces**.

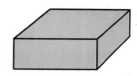 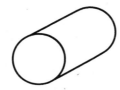

Sorting Plane Figures

Trace each side. `blue` ▷

Draw a ◯ on each corner. `red` ▷

Write how many sides and corners. **Check children's drawings.**

1.

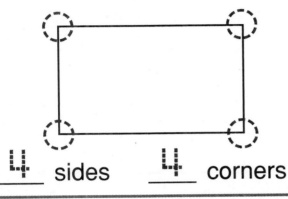

__4__ sides   __4__ corners

2.

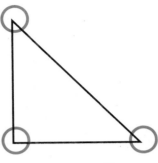

__3__ sides   __3__ corners

3.

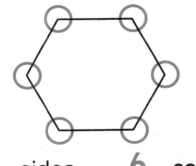

__6__ sides   __6__ corners

4.

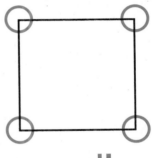

__4__ sides   __4__ corners

▶ **Problem Solving**

5. Circle the figure that has 4 corners and 4 sides.

6. Circle the figure that has 5 corners and 5 sides.

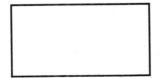   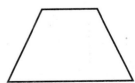

Congruence

Color the figures that are the same size and shape.

1.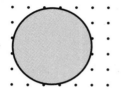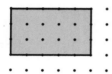

2.

3.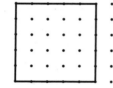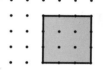

4.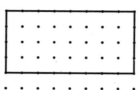

▶ **Problem Solving**

Circle the figure.

5. I have no corners.
   I have no sides.
   What am I?

 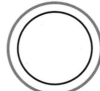 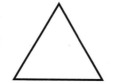

6. I have 4 corners.
   I have 4 sides.
   What am I?

Symmetry

Draw a line to make two sides that match.

**Check lines
of symmetry.**

1.
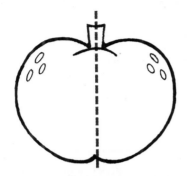

2.
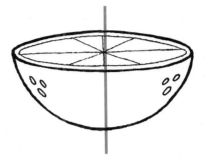

3.
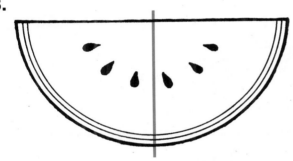

4.

5.

6.

▶ **Problem Solving**

7. Circle the figure with two parts that do not match.

Name _____

 LESSON 9.1

Open and Closed

▶ **Vocabulary**

1. Circle the **open** figure.

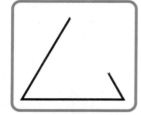 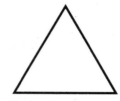

2. Circle the **closed** figure.

Color each closed figure. Circle each open figure.

3.

4.
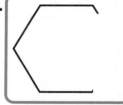 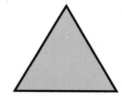 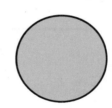 

5.
  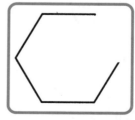 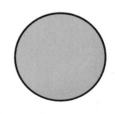

▶ **Problem Solving**

Circle the letters that are open figures.

6.

(E) B (R) (S) D (C)

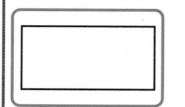

Inside, Outside, On

▶ **Vocabulary**

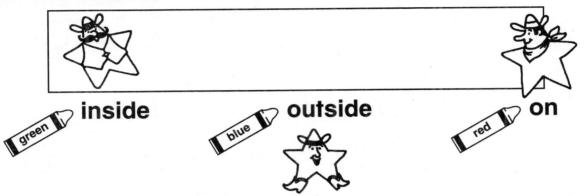

green inside   blue outside   red on

Color the squares.

1.
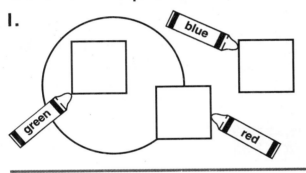

2.
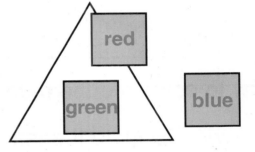

3.
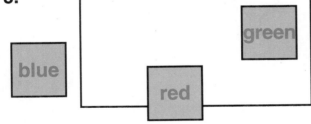

4.
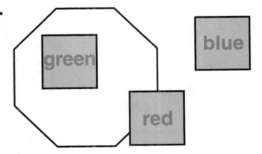

▶ **Problem Solving**

5. Which shape is inside both circles? Color the shape.

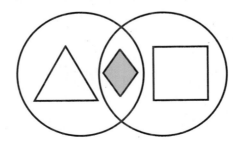

ON MY OWN   P53

Problem Solving • Draw a Picture

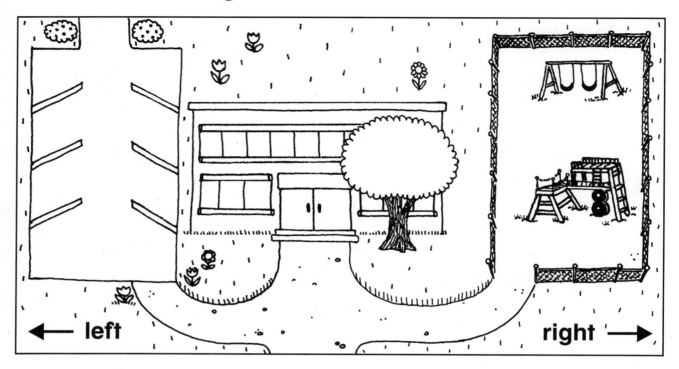

Draw to complete the map.   **Check children's drawings.**

1. Draw a 🚗 to the left of the 🏢 .

2. Draw a 📫 to the right of the 🌳 .

3. Draw two 🧒 in the ▨▨ .

Circle **left** or **right**.

4. You walk to the 🚗

   from the 🏢 .

   Which way are you going?

   (left)   right

5. You go from the 🌳

   to the 📫 .

   Which way are you going?

   left   (right)

Harcourt Brace School Publishers

Positions on a Grid

Start at ☆. Follow directions to draw shapes on the grid.

1.

Right	Up	Draw
5	6	◯
2	7	▭

Right	Up	Draw
4	3	△
1	4	▭

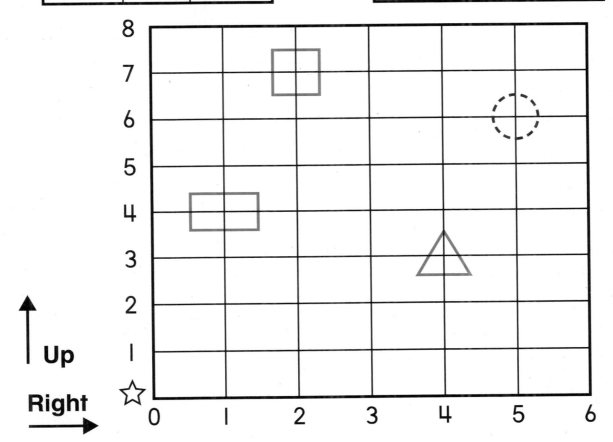

▶ Problem Solving

Look at the grid. Circle the correct shape.

2. What shape is to the
   left of the triangle?

 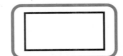

3. Which shape is farther
   to the right?

Identifying Patterns

Color the R stars ▐ red ▷ .
Color the B stars ▐ blue ▷ .
Color the Y stars ▐ yellow ▷ .
Read the pattern. Then color to continue it.

I.

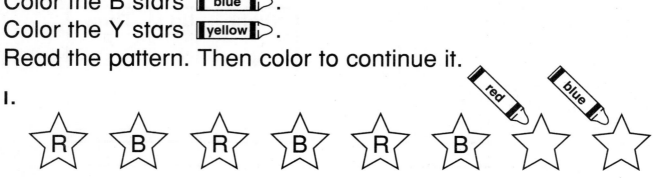

2.

3.

4.

▶ **Problem Solving**

5. Tina drew these shapes.

Alan drew these shapes.

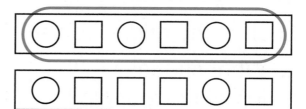

Circle the shapes that show a pattern.

Reproducing and Extending Patterns

Color the squares to copy and continue the pattern.

1.

red	blue	red	blue	red	blue

red	blue	red	blue	red	blue		

2.

blue	red	blue	red	blue	red

blue	red	blue	red	blue	red	blue	red

3.

red	red	blue	red	red	blue

red	red	blue	red	red	blue	red	red	blue

▶ **Problem Solving**   Answers will vary.

4. Make your own pattern.
Use red and blue crayons.

5. Write a number to
continue the pattern.

3	4	5	3	4	5	3

Harcourt Brace School Publishers

Making and Extending Patterns

Use shapes to continue the pattern.   **Check children's work.**
Then use the same shapes to make a different pattern.
Draw the shapes to show your new pattern.

1.

2.

3.

▶ **Problem Solving**

Draw the missing shapes to complete the pattern.

4.

5.

Harcourt Brace School Publishers

Analyzing Patterns

Find the mistake in the pattern. Cross it out.
Then use shapes to show the pattern the correct way.
Draw and color the shapes.

▶ **Problem Solving**

5. Write the numbers to continue the pattern.
   0 1 2 0 1 2 0 1 2 0 1 2 __0__ __1__ __2__

6. Use 6, 7, and 8 to make your own number pattern.

   Answers will vary.

   ___ ___ ___ ___ ___ ___ ___ ___ ___

Counting On to 12

Circle the greater number. Count on to add.

1.

$\begin{array}{r} 1 \\ + \textcircled{8} \\ \hline 9 \end{array}$
$\quad$
$\begin{array}{r} \textcircled{9} \\ + 2 \\ \hline 11 \end{array}$
$\quad$
$\begin{array}{r} \textcircled{6} \\ + 3 \\ \hline 9 \end{array}$
$\quad$
$\begin{array}{r} \textcircled{3} \\ + 2 \\ \hline 5 \end{array}$
$\quad$
$\begin{array}{r} \textcircled{7} \\ + 1 \\ \hline 8 \end{array}$

2.

$\begin{array}{r} \textcircled{9} \\ + 1 \\ \hline 10 \end{array}$
$\quad$
$\begin{array}{r} 2 \\ + \textcircled{4} \\ \hline 6 \end{array}$
$\quad$
$\begin{array}{r} \textcircled{8} \\ + 2 \\ \hline 10 \end{array}$
$\quad$
$\begin{array}{r} \textcircled{5} \\ + 3 \\ \hline 8 \end{array}$
$\quad$
$\begin{array}{r} \textcircled{9} \\ + 3 \\ \hline 12 \end{array}$

3.

$\begin{array}{r} \textcircled{8} \\ + 3 \\ \hline 11 \end{array}$
$\quad$
$\begin{array}{r} 3 \\ + \textcircled{9} \\ \hline 12 \end{array}$
$\quad$
$\begin{array}{r} 2 \\ + \textcircled{5} \\ \hline 7 \end{array}$
$\quad$
$\begin{array}{r} \textcircled{3} \\ + 2 \\ \hline 5 \end{array}$
$\quad$
$\begin{array}{r} 1 \\ + \textcircled{9} \\ \hline 10 \end{array}$

4. $\quad 2 + \textcircled{7} = \underline{9} \qquad \textcircled{7} + 3 = \underline{10} \qquad \textcircled{8} + 2 = \underline{10}$

5. $\quad \textcircled{5} + 2 = \underline{7} \qquad 3 + \textcircled{6} = \underline{9} \qquad 2 + \textcircled{3} = \underline{5}$

6. $\quad 3 + \textcircled{9} = \underline{12} \qquad \textcircled{4} + 3 = \underline{7} \qquad 1 + \textcircled{7} = \underline{8}$

▶ **Problem Solving**

7. Martha had 5 dolls. She got 2 more. How many does she have in all? $\underline{7}$ dolls

Check children's work.

Doubles to 12

Write the sums. Circle the doubles.

1.

$(4 + 4) =$ __8__ $6 + 4 =$ __10__ $(5 + 5) =$ __10__

2.

$6 + 2 =$ __8__ $(6 + 6) =$ __12__ $7 + 1 =$ __8__

3.

$7 + 2 =$ __9__ $(3 + 3) =$ __6__ $5 + 2 =$ __7__

4.

$$\begin{array}{r} 5 \\ + 3 \\ \hline 8 \end{array} \qquad \begin{array}{r} (5 \\ + 5) \\ \hline 10 \end{array} \qquad \begin{array}{r} 4 \\ + 2 \\ \hline 6 \end{array} \qquad \begin{array}{r} (2 \\ + 2) \\ \hline 4 \end{array} \qquad \begin{array}{r} 9 \\ + 2 \\ \hline 11 \end{array}$$

5.

$$\begin{array}{r} (3 \\ + 3) \\ \hline 6 \end{array} \qquad \begin{array}{r} 8 \\ + 2 \\ \hline 10 \end{array} \qquad \begin{array}{r} (6 \\ + 6) \\ \hline 12 \end{array} \qquad \begin{array}{r} 3 \\ + 6 \\ \hline 9 \end{array} \qquad \begin{array}{r} (4 \\ + 4) \\ \hline 8 \end{array}$$

▶ **Problem Solving**

Check children's work.

6. Caleb spent 6¢.
John spent 6¢.
How much did
they spend in all? __12__ ¢

Harcourt Brace School Publishers

Three Addends

Use cubes. Write the sum.

1.

$$\begin{array}{r} 4 \\ 5 \\ +\ 0 \\ \hline 9 \end{array}$$
$$\begin{array}{r} 4 \\ 3 \\ +\ 1 \\ \hline 8 \end{array}$$
$$\begin{array}{r} 6 \\ 6 \\ +\ 0 \\ \hline 12 \end{array}$$
$$\begin{array}{r} 4 \\ 5 \\ +\ 2 \\ \hline 11 \end{array}$$
$$\begin{array}{r} 6 \\ 1 \\ +\ 4 \\ \hline 11 \end{array}$$

2.

$$\begin{array}{r} 5 \\ 3 \\ +\ 2 \\ \hline 10 \end{array}$$
$$\begin{array}{r} 3 \\ 2 \\ +\ 3 \\ \hline 8 \end{array}$$
$$\begin{array}{r} 3 \\ 4 \\ +\ 3 \\ \hline 10 \end{array}$$
$$\begin{array}{r} 2 \\ 5 \\ +\ 5 \\ \hline 12 \end{array}$$
$$\begin{array}{r} 2 \\ 3 \\ +\ 4 \\ \hline 9 \end{array}$$

3.

$$\begin{array}{r} 2 \\ 1 \\ +\ 6 \\ \hline 9 \end{array}$$
$$\begin{array}{r} 3 \\ 3 \\ +\ 4 \\ \hline 10 \end{array}$$
$$\begin{array}{r} 3 \\ 5 \\ +\ 2 \\ \hline 10 \end{array}$$
$$\begin{array}{r} 1 \\ 1 \\ +\ 6 \\ \hline 8 \end{array}$$
$$\begin{array}{r} 2 \\ 5 \\ +\ 1 \\ \hline 8 \end{array}$$

4.

$$\begin{array}{r} 7 \\ 3 \\ +\ 2 \\ \hline 12 \end{array}$$
$$\begin{array}{r} 4 \\ 2 \\ +\ 2 \\ \hline 8 \end{array}$$
$$\begin{array}{r} 1 \\ 3 \\ +\ 2 \\ \hline 6 \end{array}$$
$$\begin{array}{r} 7 \\ 4 \\ +\ 1 \\ \hline 12 \end{array}$$
$$\begin{array}{r} 5 \\ 2 \\ +\ 1 \\ \hline 8 \end{array}$$

▶ **Problem Solving**

Circle the addition sentence that you think
has the greater sum. Solve to check.

5. $2 + 4 + 1 = \underline{7}$
    $\boxed{5 + 1 + 2 = \underline{8}}$

6. $\boxed{4 + 4 + 3 = \underline{11}}$
    $5 + 0 + 5 = \underline{10}$

Practice the Facts

Add. Color green the trees that have a sum of 10, 11, or 12.

1.

$$\begin{array}{r} 4 \\ 7 \\ +\ 1 \\ \hline 12 \end{array}$$ green $$\begin{array}{r} 7 \\ +\ 2 \\ \hline 9 \end{array}$$ $$\begin{array}{r} 2 \\ 5 \\ +\ 2 \\ \hline 9 \end{array}$$ $$\begin{array}{r} 3 \\ +\ 5 \\ \hline 8 \end{array}$$ $$\begin{array}{r} 6 \\ +\ 6 \\ \hline 12 \end{array}$$

2.

$$\begin{array}{r} 4 \\ 3 \\ +\ 4 \\ \hline 11 \end{array}$$ $$\begin{array}{r} 5 \\ +\ 5 \\ \hline 10 \end{array}$$ $$\begin{array}{r} 1 \\ 2 \\ +\ 3 \\ \hline 6 \end{array}$$ $$\begin{array}{r} 3 \\ 1 \\ +\ 6 \\ \hline 10 \end{array}$$ $$\begin{array}{r} 4 \\ +\ 5 \\ \hline 9 \end{array}$$

3.

$$\begin{array}{r} 2 \\ +\ 5 \\ \hline 7 \end{array}$$ $$\begin{array}{r} 8 \\ +\ 4 \\ \hline 12 \end{array}$$ $$\begin{array}{r} 5 \\ +\ 6 \\ \hline 11 \end{array}$$ $$\begin{array}{r} 2 \\ 7 \\ +\ 3 \\ \hline 12 \end{array}$$ $$\begin{array}{r} 6 \\ +\ 3 \\ \hline 9 \end{array}$$

4.

$$\begin{array}{r} 7 \\ +\ 3 \\ \hline 10 \end{array}$$ $$\begin{array}{r} 8 \\ +\ 1 \\ \hline 9 \end{array}$$ $$\begin{array}{r} 6 \\ +\ 5 \\ \hline 11 \end{array}$$ $$\begin{array}{r} 3 \\ 4 \\ +\ 3 \\ \hline 10 \end{array}$$ $$\begin{array}{r} 7 \\ 2 \\ +\ 2 \\ \hline 11 \end{array}$$

▶ **Problem Solving**

Circle the addition sentences that are correct.

5. $6 + 6 = 12$ $5 + 4 = 10$

$3 + 8 = 12$ $7 + 2 = 9$

Problem Solving • Act It Out

Act it out. Write the number sentence.

1.

5 ducks swam in the pond.
4 more ducks came to swim.
How many were swimming?

$\underline{5} + \underline{4} = \underline{9}$

$\underline{9}$ ducks

2.

6 cats played on the rug.
6 cats played on the bed.
How many were playing?

$\underline{6} + \underline{6} = \underline{12}$

$\underline{12}$ cats

3.

3 goats ran on the hill.
4 goats ran in the field.
How many were running?

$\underline{3} + \underline{4} = \underline{7}$

$\underline{7}$ goats

4.

7 white rabbits were eating.
3 brown rabbits were eating.
How many were eating?

$\underline{7} + \underline{3} = \underline{10}$

$\underline{10}$ rabbits

5.

3 frogs slept on a log.
8 frogs slept on a rock.
How many were sleeping?

$\underline{3} + \underline{8} = \underline{11}$

$\underline{11}$ frogs

6.

4 squirrels sat on a fence.
4 more squirrels came to sit.
How many were sitting?

$\underline{4} + \underline{4} = \underline{8}$

$\underline{8}$ squirrels

7. Which number sentence goes with the story? Circle it.

3 turtles were walking.
4 turtles were sleeping.
2 turtles were swimming.
How many turtles were there in all?

$\boxed{3 + 4 + 2 = 9}$

$3 + 4 + 1 = 8$

$5 + 2 + 2 = 9$

Relating Addition and Subtraction

Add. Then subtract.

1.

$8 + 4 = \underline{12}$

$12 - 4 = \underline{8}$

2.

$7 + 6 = \underline{13}$

$13 - 6 = \underline{7}$

3.

$\begin{array}{r} 6 \\ +4 \\ \hline 10 \end{array}$	$\begin{array}{r} 10 \\ -4 \\ \hline 6 \end{array}$	$\begin{array}{r} 7 \\ +2 \\ \hline 9 \end{array}$	$\begin{array}{r} 9 \\ -2 \\ \hline 7 \end{array}$

$\begin{array}{r} 9 \\ +1 \\ \hline 10 \end{array}$ $\begin{array}{r} 10 \\ -1 \\ \hline 9 \end{array}$

4.

$\begin{array}{r} 5 \\ +3 \\ \hline 8 \end{array}$ $\begin{array}{r} 8 \\ -3 \\ \hline 5 \end{array}$ $\begin{array}{r} 9 \\ +2 \\ \hline 11 \end{array}$ $\begin{array}{r} 11 \\ -2 \\ \hline 9 \end{array}$ $\begin{array}{r} 8 \\ +2 \\ \hline 10 \end{array}$ $\begin{array}{r} 10 \\ -2 \\ \hline 8 \end{array}$

▶ Problem Solving

5. Two numbers are added. The sum is 8.
One number is 5. What is the other number? ___3___

6. Two numbers are added. The sum is 6.
One number is 4. What is the other number? ___2___

Counting Back

▶ **Vocabulary**

Count back to subtract.
Use the number line if you need it.

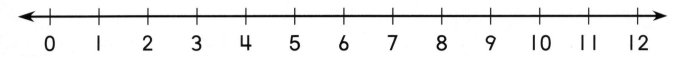

1.

$$\begin{array}{r} 11 \\ -\ 3 \\ \hline 8 \end{array}$$
$$\begin{array}{r} 9 \\ -\ 1 \\ \hline 8 \end{array}$$
$$\begin{array}{r} 12 \\ -\ 2 \\ \hline 10 \end{array}$$
$$\begin{array}{r} 7 \\ -\ 2 \\ \hline 5 \end{array}$$
$$\begin{array}{r} 6 \\ -\ 1 \\ \hline 5 \end{array}$$

2.

$$\begin{array}{r} 10 \\ -\ 2 \\ \hline 8 \end{array}$$
$$\begin{array}{r} 7 \\ -\ 1 \\ \hline 6 \end{array}$$
$$\begin{array}{r} 9 \\ -\ 3 \\ \hline 6 \end{array}$$
$$\begin{array}{r} 8 \\ -\ 1 \\ \hline 7 \end{array}$$
$$\begin{array}{r} 12 \\ -\ 3 \\ \hline 9 \end{array}$$

3.

$$\begin{array}{r} 8 \\ -\ 2 \\ \hline 6 \end{array}$$
$$\begin{array}{r} 11 \\ -\ 2 \\ \hline 9 \end{array}$$
$$\begin{array}{r} 5 \\ -\ 1 \\ \hline 4 \end{array}$$
$$\begin{array}{r} 4 \\ -\ 3 \\ \hline 1 \end{array}$$
$$\begin{array}{r} 10 \\ -\ 3 \\ \hline 7 \end{array}$$

▶ **Problem Solving**

4. A number line was painted on the playground.
A bird was standing on number 9.
It took 3 hops back. What number was it on? ___6___

Harcourt Brace School Publishers

Compare to Subtract

Compare. Then subtract.

1.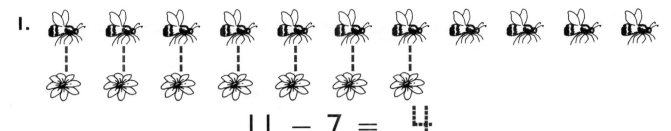

$11 - 7 = \underline{4}$

2.

$10 - 8 = \underline{2}$

3.

$11 - 9 = \underline{2}$

4.

$12 - 8 = \underline{4}$

▶ **Problem Solving**

5. Ashley has 12 flowers. Cody has 9 flowers. How many fewer flowers does Cody have?

$\underline{3}$ fewer flowers

Check children's work.

Fact Families

Add and subtract.

1.

$6 + 4 = 10$
$4 + 6 = 10$
$10 - 4 = 6$
$10 - 6 = 4$

2.

$3 + 4 = 7$
$4 + 3 = 7$
$7 - 4 = 3$
$7 - 3 = 4$

3.

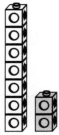

$7 + 2 = 9$
$2 + 7 = 9$
$9 - 2 = 7$
$9 - 7 = 2$

4.

$5 + 6 = 11$
$6 + 5 = 11$
$11 - 6 = 5$
$11 - 5 = 6$

5.

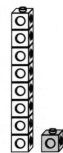

$8 + 1 = 9$
$1 + 8 = 9$
$9 - 1 = 8$
$9 - 8 = 1$

6.

$4 + 2 = 6$
$2 + 4 = 6$
$6 - 2 = 4$
$6 - 4 = 2$

▶ **Problem Solving**

7. Write or draw a story problem that uses these numbers.

   7   3   10

Write the number sentence your story shows.

____ ◯ ____ = ____

Check children's work.

Possible answers:
$7 + 3 = 10$ $3 + 7 = 10$
$10 - 3 = 7$ $10 - 7 = 3$

Harcourt Brace School Publishers

Problem Solving • Write a Number Sentence

Write the number sentence the story problem shows.

1. There are 10 dogs.
There are 7 cats.
How many more dogs
than cats are there?

 ___3___ more dogs

$$10 \ominus 7 = 3$$

2. There are 9 goldfish.
There are 7 guppies.
How many more goldfish
than guppies are there?

 ___2___ more goldfish

$$9 \ominus 7 = 2$$

3. Jenny had 6 cookies.
She ate 3 of them.
How many cookies
does she have left?

 ___3___ cookies

$$6 \ominus 3 = 3$$

4. Jason had 2 apples.
His grandmother gave
him 3 more apples.
How many apples does
he have in all?

 ___5___ apples

$$2 \oplus 3 = 5$$

Tens

Write how many tens. Write the number.

1.

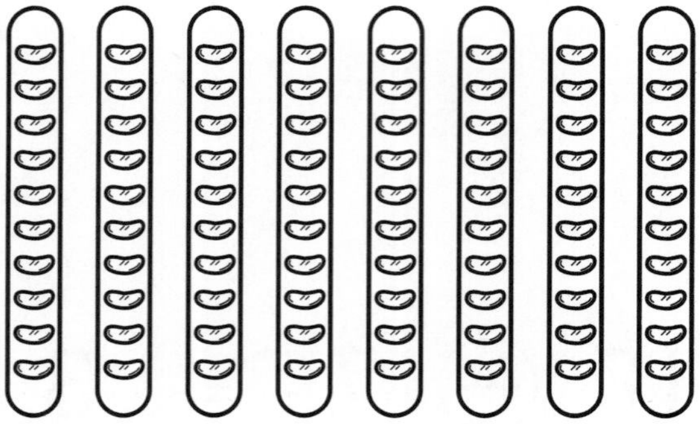

____8____ tens = __80__
eighty

2.

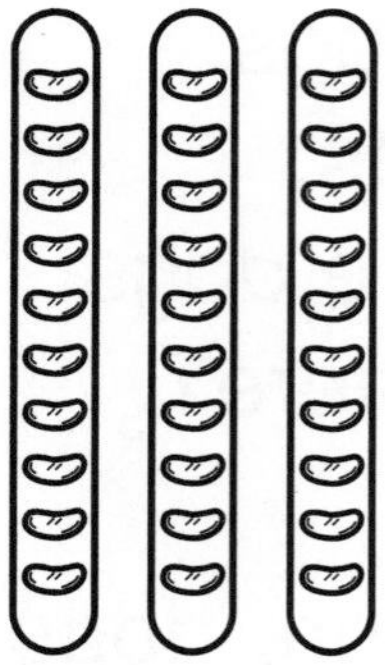

____3____ tens = __30__
thirty

3.

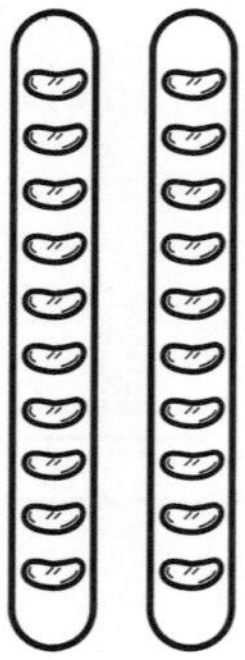

____2____ tens = __20__
twenty

4.

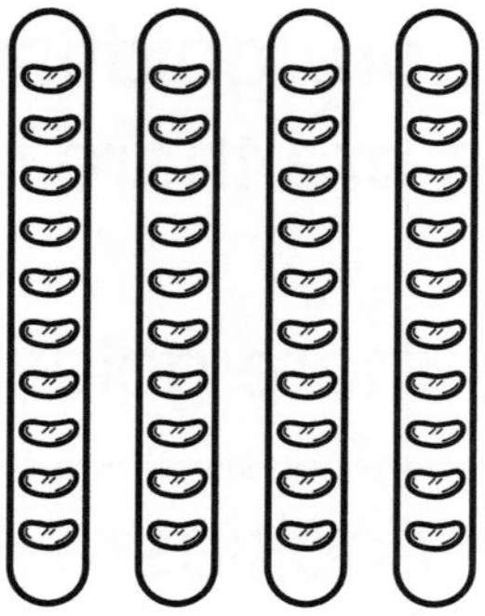

____4____ tens = __40__
forty

▶ **Problem Solving**

5. Circle the box that has 60.

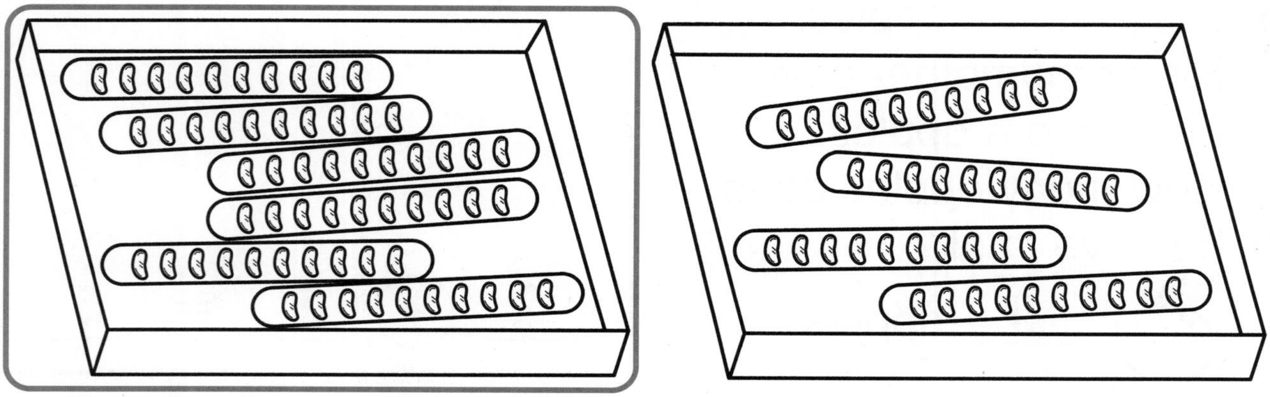

Harcourt Brace School Publishers

Tens and Ones to 20

▶ Vocabulary

1. Circle the **tens**.

2. Circle the **ones**.

Write how many tens and ones. Write the number.

3. _____1_____ ten _____5_____ ones = ___15___

4. _____1_____ ten _____2_____ ones = ___12___

5. _____2_____ tens _____0_____ ones = ___20___

6. _____1_____ ten _____7_____ ones = ___17___

▶ Problem Solving

7. Joe has 10 toy cars.
Mark gives him 2 more.
How many toy cars does
Joe have in all?

___12___ cars

Check children's work.

Tens and Ones to 50

Write how many.

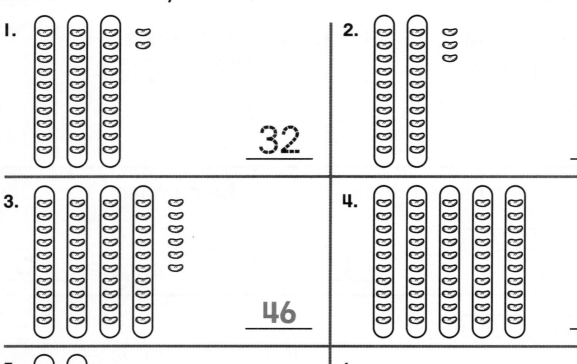

1. 32

2. 23

3. 46

4. 50

5. 28

6. 9

▶ **Problem Solving**

Circle the box that has more.

7.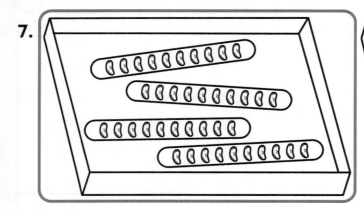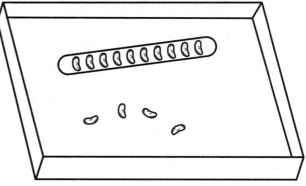

Harcourt Brace School Publishers

Tens and Ones to 80

Write the number.

1. 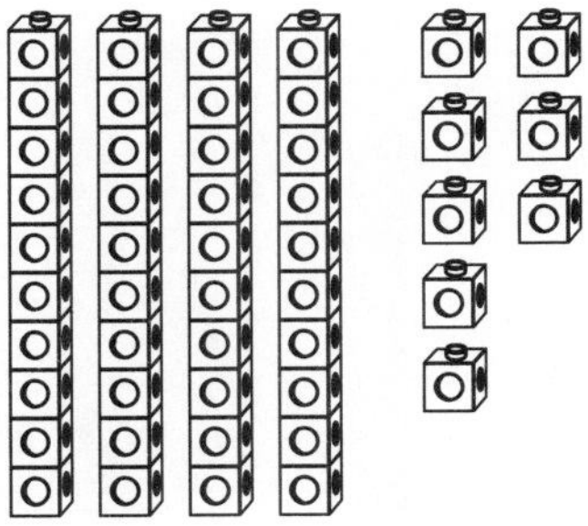 48

2. 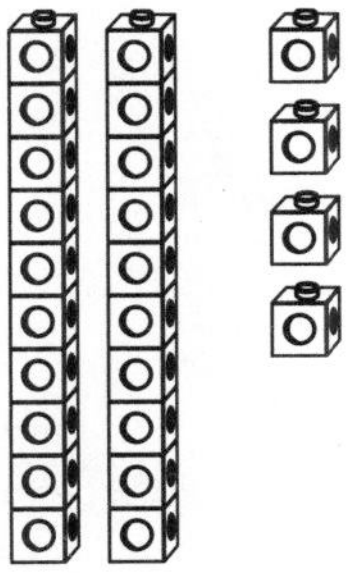 24

3. 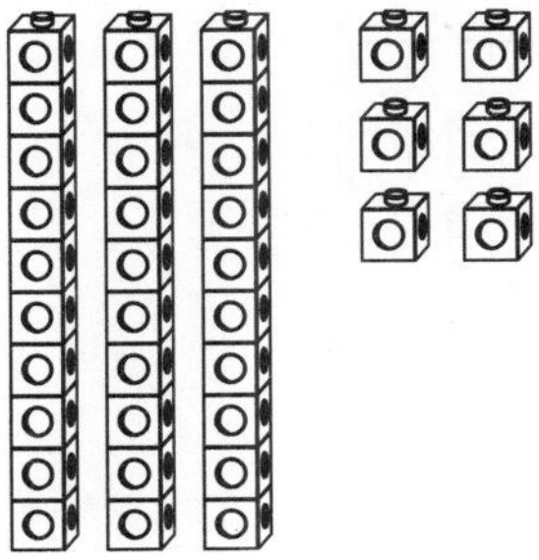 36

4. 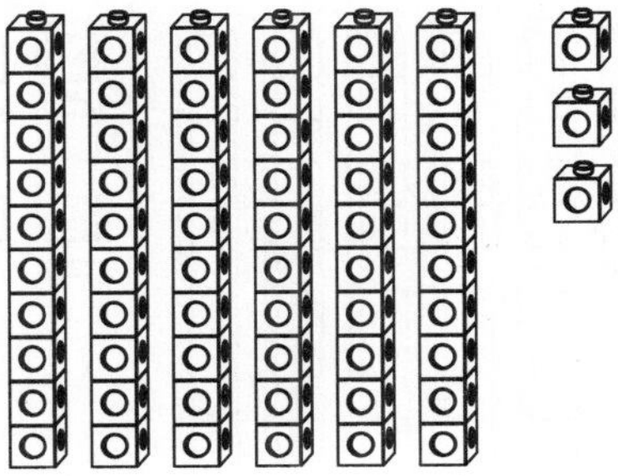 63

5. 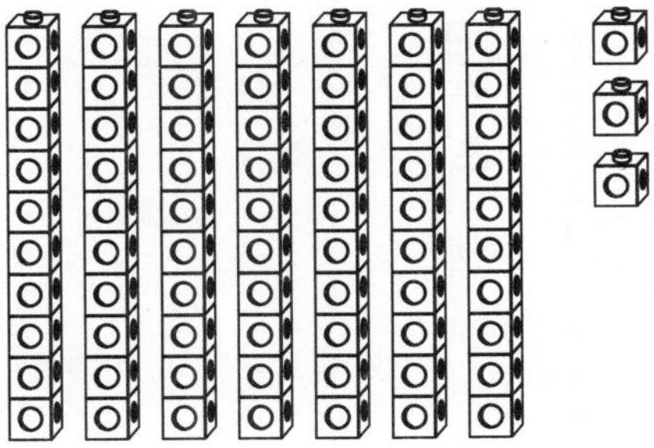 73

6. 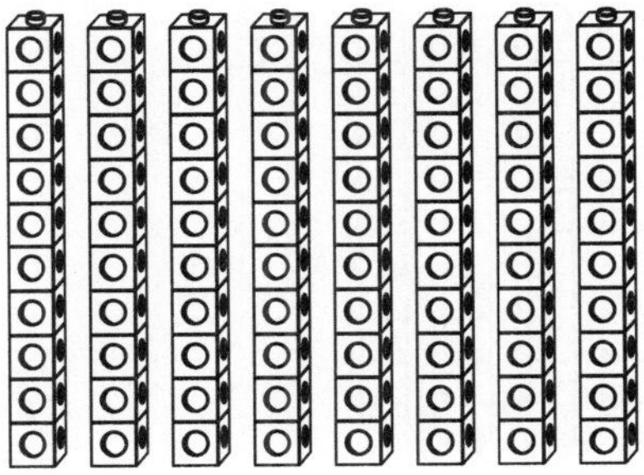 80

▶ **Problem Solving**

Write the number.

7. Mary has I ten and 3 ones.
Joe has 3 tens and I one.
Write the number that tells
how many each one has.

Mary __13__   Joe __31__

Check children's work.

Tens and Ones to 100

Write the number.

1. 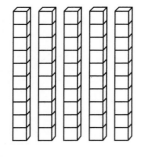 5 1

2. 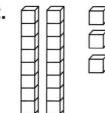 23

3. 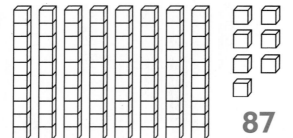 87

4. 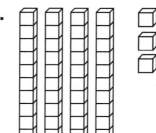 45

5. 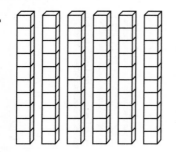 6 1

6. 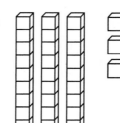 33

▶ **Problem Solving**

Draw a picture to solve.

7. Kenda picked apples. She had 3 groups of 10 apples and 3 left over. How many apples did she have in all?

____33____ apples

Check children's work.

Harcourt Brace School Publishers

Estimating 10

Circle the better estimate.

1.

(more than 10)

fewer than 10

2.

more than 10

(fewer than 10)

3.

more than 10

(fewer than 10)

4.

(more than 10)

fewer than 10

▶ **Problem Solving**

5. Matt has more than 15 but fewer than 20 oranges. Write the numbers that tell how many oranges Matt could have.

16   17   18   19

Ordinals

Match.

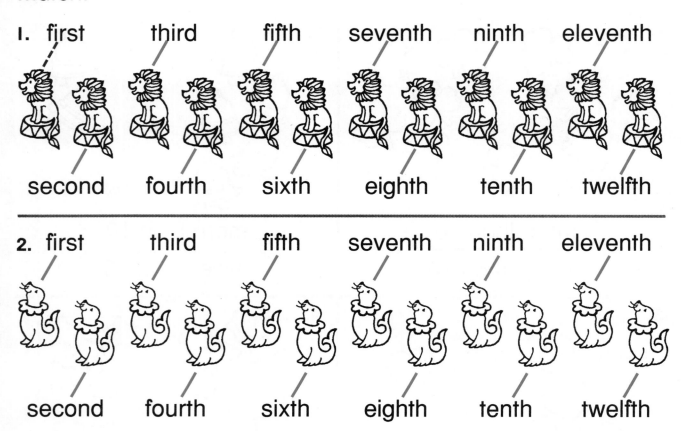

1. first   third   fifth   seventh   ninth   eleventh

    second   fourth   sixth   eighth   tenth   twelfth

2. first   third   fifth   seventh   ninth   eleventh

    second   fourth   sixth   eighth   tenth   twelfth

▶ **Problem Solving**

Circle the answer.

3. Which animal is first?

    (rabbit)   dog   duck   turtle

4. Which animal is third?

    rabbit   dog   (duck)   turtle

Harcourt Brace School Publishers

Greater Than

▶ **Vocabulary**

Write the numbers. Circle the **greater** number.

1.

(50)

25

2.

85

(99)

3.

(83)

62

4.

19

(70)

▶ **Problem Solving**

5. Bill has 2 dogs. Marc has more dogs than Bill.
Circle the boy that is Marc.

6. The bus has 21 people on it. The plane
has 40 people on it. Circle the one
with the greater number on it.

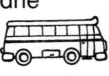

Less Than

▶ Vocabulary

Write the numbers. Circle the number that is **less.**

I.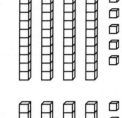

45

(43)

2.

(25)

30

3.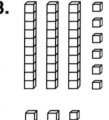

36

(22)

4.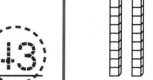

54

(35)

▶ Problem Solving

5. Circle the one who has the most pennies.

Jess   Sue   ✗Pat   (Bill)

6. Mark an **X** on the one who has fewer pennies than Jess.

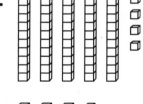

Harcourt Brace School Publishers

Before, After, Between

▶ Vocabulary

 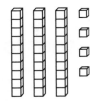 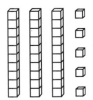

| **33** is just **before 34.** | **34** is **between 33** and **35.** | **36** is just **after 35.** |

Complete the tables.

	before		after
1.	44	45	46
2.	59	60	61
3.	86	87	88
4.	22	23	24
5.	58	59	60
6.	10	11	12
7.	35	36	37
8.	97	98	99
9.	69	70	71

		between	
10.	46	47	48
11.	83	84	85
12.	19	20	21
13.	60	61	62
14.	55	56	57
15.	97	98	99
16.	78	79	80
17.	25	26	27
18.	89	90	91

▶ Problem Solving

Write the numbers in order.

19.	25	26	24
	24	25	26

20.	55	53	54
	53	54	55

Order to 100

▶ **Vocabulary**

Write the numbers.
Then write them in order from **least** to **greatest.**

1.

8   28   15   38

_8_   _15_   _28_   _38_

Write the numbers in order from least to greatest.

2. 57   59   53   55
    53   55   57   59

3. 32   52   62   42
    32   42   52   62

4. 22   40   39   54
    22   39   40   54

5. 61   58   41   73
    41   58   61   73

▶ **Problem Solving**

6. Write three numbers that are
   between 52 and 57
   and that come after 53.   _54_   _55_   _56_

Harcourt Brace School Publishers

Counting by Tens

Count by tens. Write how many.

1.

<u>10</u>   20   30   40   50   60

2.

10   20   30   40   50   60   70

▶ **Problem Solving**

Count by tens to fill in the table.
Use the table to answer the questions.

Megan has 20 pennies on Sunday.
She saves 10 pennies every day for one week.

Sunday	Monday	Tuesday	Wednesday	Thursday	Friday	Saturday
20	30	40	50	60	70	80

3. How many pennies does she have on Tuesday?

<u>40</u> pennies

4. How many pennies does she have on Friday?

<u>70</u> pennies

5. How many pennies does she have on Saturday?

<u>80</u> pennies

Name _____

Counting by Fives

1. Write the missing numbers.   **Check children's coloring.**
   Count by fives. Color those boxes red.
   Count by tens. Color those boxes blue.

1	2	3	4	5 red	6	7	8	9	10
11	12	13	14	15 red	16	17	18	19	red 20 blue
21	22	23	24	25 red	26	27	28	29	red 30 blue
31	32	33	34	35 red	36	37	38	39	red 40 blue
41	42	43	44	45 red	46	47	48	49	red 50 blue
51	52	53	54	55 red	56	57	58	59	red 60 blue
61	62	63	64	65 red	66	67	68	69	red 70 blue
71	72	73	74	75 red	76	77	78	79	red 80 blue
81	82	83	84	85 red	86	87	88	89	red 90 blue
91	92	93	94	95 red	96	97	98	99	red 100 blue

▶ **Problem Solving**

2. Circle the numbers that are greater than 50.

(80)   30   (90)   40   20

Counting by Twos

Count by twos. Write the missing numbers.

1.

1	2	3	4	5	6	7	8	9	10
11	12	13	14	15	16	17	18	19	20
21	22	23	24	25	26	27	28	29	30
31	32	33	34	35	36	37	38	39	40
41	42	43	44	45	46	47	48	49	50
51	52	53	54	55	56	57	58	59	60
61	62	63	64	65	66	67	68	69	70
71	72	73	74	75	76	77	78	79	80
81	82	83	84	85	86	87	88	89	90
91	92	93	94	95	96	97	98	99	100

2.
My number is between 40 and 50. It is 2 more than 45. What is my number?

47

3.
My number is between 80 and 90. It is 2 more than 87. What is my number?

89

Even and Odd Numbers

▶ Vocabulary

Circle **even** or **odd**.

1.
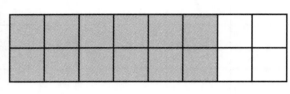

12   (even)   odd

2.
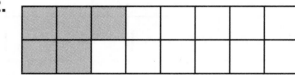

5   even   (odd)

Color the squares to show each number. Circle **even** or **odd**.

3.

7   even   (odd)

4.
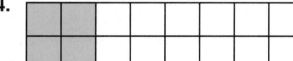

4   (even)   odd

5.

14   (even)   odd

6.
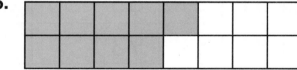

9   even   (odd)

▶ Problem Solving

Circle **even** or **odd**.

7. Ann has 2 .
Each has 2 .
Does she have an even or
odd number of ?

(even)   odd

Pennies and Nickels

▶ **Vocabulary**

1. Circle the **penny.**
2. Cross out the **nickel.**

Count. Write the amount.

3.

  _____

__1__ ¢, __2__ ¢, __3__ ¢ | 3 | ¢

4.

__1__ ¢, __2__ ¢, __3__ ¢, __4__ ¢, __5__ ¢, __6__ ¢ | 6 | ¢

5.

__5__ ¢, __10__ ¢ | 10 | ¢

6.

__5__ ¢, __10__ ¢, __15__ ¢ | 15 | ¢

▶ **Problem Solving**

Circle the greater amount.

7.

Pennies and Dimes

▶ Vocabulary

1. Circle the **penny**.
2. Cross out the **dime**.

Count by tens. Write the amount.

3. <u>10</u> ¢, <u>20</u> ¢   [20] ¢

4.   <u>10</u> ¢, <u>20</u> ¢, <u>30</u> ¢, <u>40</u> ¢   [40] ¢

5.   <u>10</u> ¢, <u>20</u> ¢, <u>30</u> ¢, <u>40</u> ¢, <u>50</u> ¢   [50] ¢

6.   <u>10</u> ¢, <u>20</u> ¢, <u>30</u> ¢   [30] ¢

7.   <u>10</u> ¢, <u>20</u> ¢, <u>30</u> ¢, <u>40</u> ¢, <u>50</u> ¢, <u>60</u> ¢, <u>70</u> ¢   [70] ¢

▶ Problem Solving

Circle the least amount.

8.

Harcourt Brace School Publishers

Name _____

Counting Collections of Nickels and Pennies

Count by fives. Then count on by ones.
Write the amount.

1.   8 ¢

2.   11 ¢

3.   22 ¢

4.   26 ¢

▶ **Problem Solving**

Mark an **X** on the greater amount.

5.

6.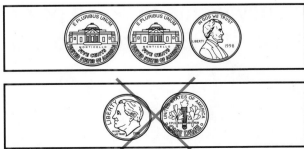

Counting Collections of Dimes and Pennies

Count by tens. Then count on by ones.
Write the amount.

1.   　$\boxed{31}$ ¢

2.   　$\boxed{22}$ ¢

3.     　$\boxed{53}$ ¢

4.   　$\boxed{41}$ ¢

5.   　$\boxed{14}$ ¢

▶ **Problem Solving**

6. Lucy wants to buy a paint set.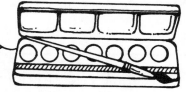
It costs 43¢. Circle the coins Lucy needs.

Harcourt Brace School Publishers

Problem Solving • Choose the Model

Which two groups in each row add up
to the amount on the tag? Color them.

1.

2.

3.

4.

▶ **Problem Solving**

5. You have

     .

Circle the toy you can buy.

Trading Pennies, Nickels, and Dimes

Trade for nickels and dimes. Use the fewest coins.
Draw how many you have.

1.

2.

2 dimes, 1 nickel

3.

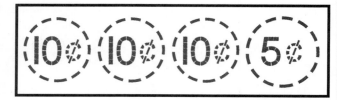

I dime, I nickel

4.

I dime

▶ **Problem Solving**

5. Wayne bought a watch.
He used 4 nickels.

Did he use the fewest
coins that equal 20¢? __no__

Show the price using
the fewest coins.

2 dimes

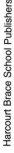

Harcourt Brace School Publishers

Equal Amounts

Show the amount in 2 ways.
Circle the way that uses the fewest coins.

Answers may vary.

I.

2.

3.

 **Problem Solving**

4. Lance needs 30¢ to buy a notebook.
Mark an **X** on the groups that do not equal 30¢.

Which way uses the fewest coins to show 30¢? Circle it.

How Much Is Needed?

Circle the coins you need. Use the fewest coins.

1.

2.

3.

4.

▶ **Problem Solving**

5. Ruben has 15 .
He wants to trade his
pennies for other coins.
Show the fewest coins
he can have.

Check children's work.

I dime, I nickel

Harcourt Brace School Publishers

Quarter

▶ **Vocabulary**

Circle the **quarter**.

Write each amount. Circle the coins that equal a .

1.

_25_ ¢

2.

_20_ ¢

3.

_26_ ¢

4.

_25_ ¢

5.

_25_ ¢

6.

_9_ ¢

▶ **Problem Solving**

7. Nathan wants to buy a .
It costs 25¢.
Circle the coins he needs.

Problem Solving • Act It Out

Play store. Work with a partner.
Use the fewest coins to buy things.
Take turns.   **Answers will vary.**

Draw what you bought.	Draw the coins you used.
1.	
2.	

Ordering Months and Days

FEBRUARY

Sunday	Monday	Tuesday	Wednesday	Thursday	Friday	Saturday
1	2	3 blue	4	5	6	7
8	9	10 blue	11	12	13	14
15	16	17 blue	18	19	20	21
22	23	24 blue	25	26	27	28

Write the days of the week in order.

1. _Sunday_

2. _Monday_

3. _Tuesday_

4. _Wednesday_

5. _Thursday_

6. _Friday_

7. _Saturday_

8. Color the Tuesdays .

▶ Problem Solving

9. The ball game is on the day before Saturday. Write the day.

   Friday

10. Lee's birthday is in the last month of the year. Circle the month.

    October   (December)

Reading the Calendar

Fill in the calendar for next month.
Use the calendar to answer the questions.

Answers may vary.

_____

Sunday	Monday	Tuesday	Wednesday	Thursday	Friday	Saturday

1. On what day does the month end?
   _____

2. What is the date of the first Sunday?
   _____

3. What is the date of the first Friday?
   _____

▶ Problem Solving

4. Mike's birthday is on May 7.
   This year May 6 is on Thursday.
   On what day of the week is Mike's birthday?

   _____
   Friday

Ordering Events

Answers will vary.

Draw something special you do on each of these days.

Saturday night
Sunday morning
Monday afternoon

▶ **Problem Solving**

1. Circle the month that comes before December.

   January　　(November)　　February

2. Write the month that comes
   after October.

   _____
   November

Estimating Time

Circle the one that takes longer to do.

1.

2.

3.

▶ **Problem Solving**

4. Tyler and Sarah live next to each other.
   Tyler rides his bike home from school.
   Sarah walks home.
   Does it take more time for
   Tyler or Sarah to get home?

   _____
   Sarah

Harcourt Brace School Publishers

Reading the Clock

Use your clock. Show the time.
Write the time two ways.

1.

___**1**___ o'clock

1:00

2.

___**5**___ o'clock

5:00

3.

___**9**___ o'clock

9:00

4.

___**3**___ o'clock

3:00

5.

___**12**___ o'clock

12:00

6.

___**4**___ o'clock

4:00

▶ **Problem Solving**

7. It is 7 o'clock. Jenny has
to go to bed in one hour.

What time does Jenny go to
bed? Write the time two ways.

___**8**___ o'clock

8:00

Name _____

Hour

Write the time.

1.

| 12:00 |

2.

| 5:00 |

3.

| 1:00 |

4.
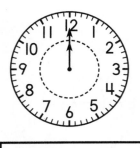

| 9:00 |

5.

| 3:00 |

6.

| 7:00 |

7.

| 2:00 |

8.

| 6:00 |

9.

| 4:00 |

▶ **Problem Solving**

10. Write the time on the clock
so that it shows 1 hour
later than 6 o'clock.

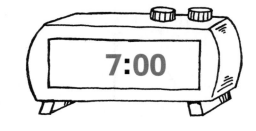

Name _____

Hour

▶ **Vocabulary**

Draw the **hour hand** and the **minute hand**.

1. `6:00`

2. `10:00`

3. `3:00`

4. `12:00`

5. `7:00`

6. `1:00`

7. `8:00`

8. `2:00`

9. `4:00`

▶ **Problem Solving**

10. Jonathan eats dinner at 6 o'clock. He goes to bed 2 hours later. Write the time he goes to bed.

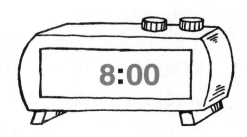

8:00

Half Hour

Write the time.

1.

5:30

2.

7:30

3.

1:30

4.

2:30

5.

9:30

6.

3:30

7.

12:30

8.

10:30

9.

6:30

▶ **Problem Solving**

10. The movie starts at 7:30.
It lasts for two hours.
What time is the movie over?
Circle the clock that shows
when the movie is over.

Problem Solving • Act It Out

About how long would it take? Circle your estimate.
Then act it out to see if you are right.

1. put 10 chairs in a circle

> (more than a minute)
> less than a minute

2. put a stamp on a letter

more than a minute
(less than a minute)

3. open a door

more than a minute
(less than a minute)

4. write 10 spelling words

(more than a minute)
less than a minute

5. read a big book

(more than a minute)
less than a minute

6. sharpen a pencil

more than a minute
(less than a minute)

Using Nonstandard Units

Estimate. Then use 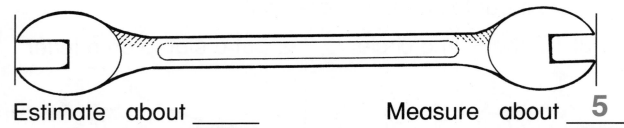 to measure.   Estimates may vary.

1.

Estimate   about _____   Measure   about **5**

2.

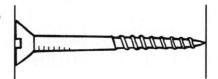

Estimate   about _____   Measure   about **2**

3.

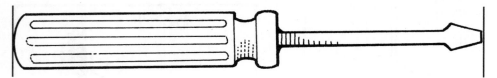

Estimate   about _____   Measure   about **4**

4.

Estimate   about _____   Measure   about **3**

▶ **Problem Solving**

5. Use and to measure.
Circle which way uses more.

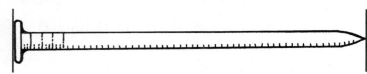

Harcourt Brace School Publishers

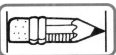

Measuring in Inch Units

 Vocabulary

1. Circle the pencil that is 1 **inch** long.

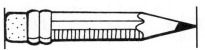

Count the inch units. Write how many inches long.

2.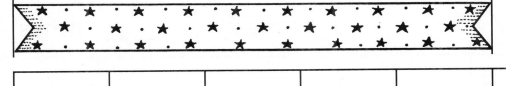

_____ 5 _____ inches

3.

_____ 4 _____ inches

4.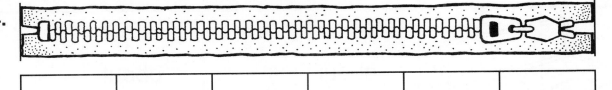

_____ 6 _____ inches

Check children's drawing.

 Problem Solving

5. One ▭ is about
1 inch long. About how
many inches long are
3 ▭? Draw a line
to show.

about ___ 3 ___ inches long

Using an Inch Ruler

You will need: objects, an inch ruler
Estimate. Then use an inch ruler to measure.

Answers will vary.

Object	Estimate	Measure
1.	about ____ inches	about ____ inches
2. TISSUES	about ____ inches	about ____ inches
3.	about ____ inches	about ____ inches
4.	about ____ inches	about ____ inches

▶ **Problem Solving**

5. Measure each chain. If you joined the two chains,
how long would the new chain be?

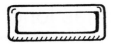  ___3___ inches

Harcourt Brace School Publishers

Name _____

Measuring in Centimeter Units

▶ **Vocabulary**

Measure. Circle the rope that is 1 **centimeter** long.

How many centimeters long?
Count the centimeter units. Write how many.

1.

_**3**_ centimeters

2.

_**4**_ centimeters

3.

_**7**_ centimeters

4.

_**5**_ centimeters

▶ **Problem Solving**   **Check children's drawings.**

5. Draw a 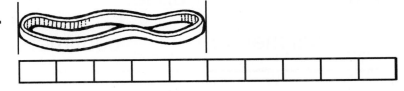 10 centimeters long.

Using a Centimeter Ruler

You will need: objects, a centimeter ruler   Answers will vary.
Estimate. Then use a centimeter ruler to measure.

Object	Estimate	Measure
I.	about ____ centimeters	about ____ centimeters
2.	about ____ centimeters	about ____ centimeters
3.	about ____ centimeters	about ____ centimeters
4.	about ____ centimeters	about ____ centimeters

▶ Problem Solving

5. Use a centimeter ruler. Measure the sides of the rectangle.

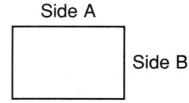

Side A

Side B

Side A __3__ centimeters

Side B __2__ centimeters

Harcourt Brace School Publishers

Using a Balance

▶ **Vocabulary**

1. Circle the object that is **heavier.**

2. Circle the object that is **lighter.**

You will need: a 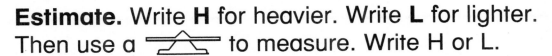. Work with a partner. Find each object. Is the object heavier or lighter than a bottle of glue?

Estimate. Write **H** for heavier. Write **L** for lighter. Then use a ⟨balance⟩ to measure. Write H or L.

Object	Estimate	Measure
3.	Answers will vary.	Some answers may vary. **L**
4.		**H**
5.		**L**

▶ **Problem Solving**

6. Ashley has 2 cups. One is full of marbles. One is full of chalk. Circle the cup that is heavier.

Using Nonstandard Units

You will need: a ▱ and ▣

Find each object. About how many ▣ does it take to balance the scale? Estimate. Then measure.

Object	Estimate	Measure
1.	Answers will vary. about _____ ▣	Answers will vary. about _____ ▣
2.	about _____ ▣	about _____ ▣
3. Crayons	about _____ ▣	about _____ ▣
4.	about _____ ▣	about _____ ▣

▶ **Problem Solving**   Check children's drawings.

Look at your measures for the objects.

5. Draw a picture of
   the heaviest object.

6. Draw a picture of
   the lightest object.

Measuring with Cups

About how many cups of rice does
each object hold? Estimate. Then measure.

Object	Estimate	Measure
I.	Answers will vary. about _____ cups	Answers will vary. about _____ cups
2.	about _____ cups	about _____ cups
3.	about _____ cups	about _____ cups
4. Quart	about _____ cups	about _____ cups

▶ **Problem Solving**   Check children's drawings.

Look at the container.
Draw a container beside it that holds more.

5.

6.

Harcourt Brace School Publishers

Temperature
Hot and Cold

Circle the picture that shows something hot.

1.

2.

Circle the picture that shows something cold.

3.

4.

▶ **Problem Solving**

Check children's drawings.

5. Draw something hot.

6. Draw something cold.

LESSON 22.1

Equal and Unequal Parts of Wholes

▶ **Vocabulary**

1. Circle the figure that shows **equal parts.**

2. Mark an **X** on the one that does not show equal parts.

  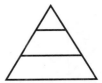

Circle the figures that show equal parts.

3.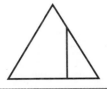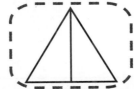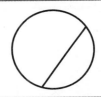

4.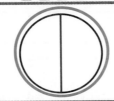

5.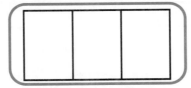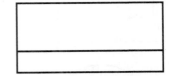

6.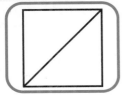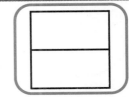

▶ **Problem Solving**

Check children's drawings. Answers may vary.

Draw lines to show where you would cut this cake.

7. Each child wants an equal share.

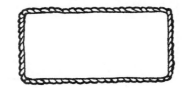 

Halves

Find the figures that show halves. Color $\frac{1}{2}$ red.

1.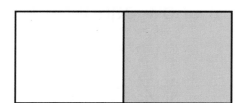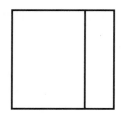

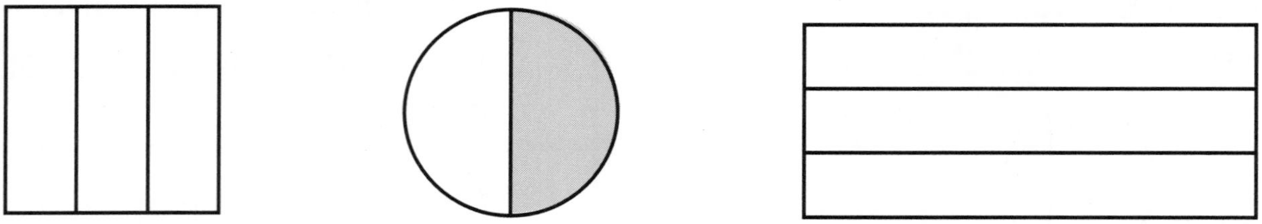

2.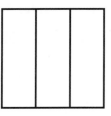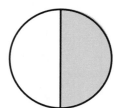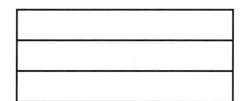

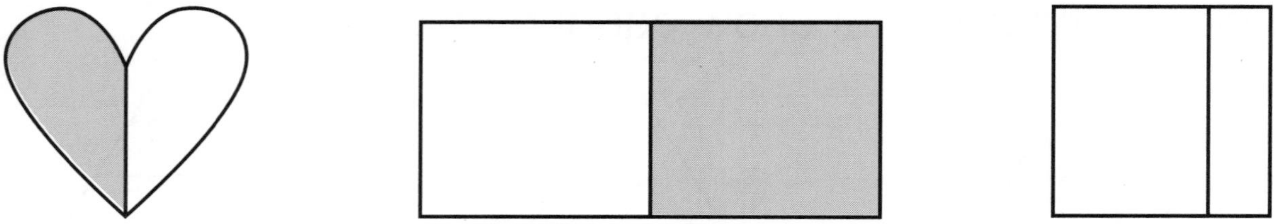

3.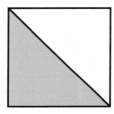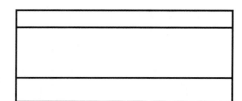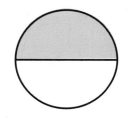

4.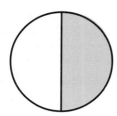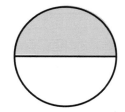

► **Problem Solving**

5. Two rabbits shared 4 carrots. Each ate one half of the 4 carrots. How many carrots did each rabbit eat?

____2____ carrots

Fourths

Check children's coloring.

Color one part 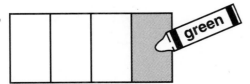 green. Circle the fraction.

1.

$\frac{1}{2}$ $\left(\frac{1}{4}\right)$

$\left(\frac{1}{2}\right)$ $\frac{1}{4}$

$\left(\frac{1}{2}\right)$ $\frac{1}{4}$

2.

$\frac{1}{2}$ $\left(\frac{1}{4}\right)$

$\left(\frac{1}{2}\right)$ $\frac{1}{4}$

$\frac{1}{2}$ $\left(\frac{1}{4}\right)$

3.

$\left(\frac{1}{2}\right)$ $\frac{1}{4}$

$\frac{1}{2}$ $\left(\frac{1}{4}\right)$

$\left(\frac{1}{2}\right)$ $\frac{1}{4}$

▶ **Problem Solving**

Draw lines. Then color one part to show the fraction.

Check children's work.

4.  $\frac{1}{2}$

5.  $\frac{1}{4}$

ON MY OWN P 115

Thirds

Color one part 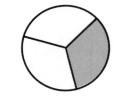. Circle the fraction.

Check children's coloring.

I.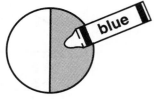

$\frac{1}{3}$ $\left(\frac{1}{2}\right)$ $\frac{1}{4}$ $\left(\frac{1}{3}\right)$ $\frac{1}{2}$ $\frac{1}{4}$ $\frac{1}{3}$ $\frac{1}{2}$ $\left(\frac{1}{4}\right)$

2.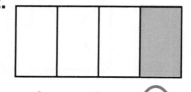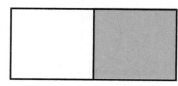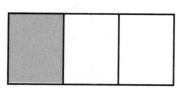

$\frac{1}{3}$ $\frac{1}{2}$ $\left(\frac{1}{4}\right)$ $\frac{1}{3}$ $\left(\frac{1}{2}\right)$ $\frac{1}{4}$ $\left(\frac{1}{3}\right)$ $\frac{1}{2}$ $\frac{1}{4}$

3.

$\left(\frac{1}{3}\right)$ $\frac{1}{2}$ $\frac{1}{4}$ $\frac{1}{3}$ $\frac{1}{2}$ $\left(\frac{1}{4}\right)$ $\frac{1}{3}$ $\left(\frac{1}{2}\right)$ $\frac{1}{4}$

▶ **Problem Solving**

What part is left? Circle the fraction.

4.  $\frac{1}{3}$ $\left(\frac{1}{2}\right)$ $\frac{1}{4}$ **5.** 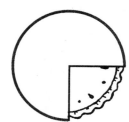 $\frac{1}{3}$ $\frac{1}{2}$ $\left(\frac{1}{4}\right)$

Harcourt Brace School Publishers

Visualizing Results

Think about sharing this pie.
Circle the picture that answers the question.

1. There are 3 children. Each gets an equal share. How would you cut the pie?

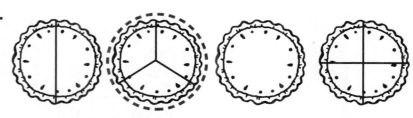

2. There are 2 children. Each gets an equal share. How would you cut the pie?

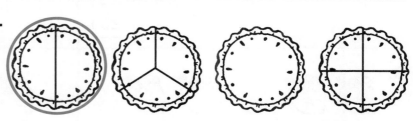

3. There are 4 children. Each gets an equal share. How much is an equal share?

▶ **Problem Solving**

Drawings may vary.

Draw lines to show the answer.

4. Robin and Rico got a small pizza. They want equal shares. How should they cut the pizza?

5. Bryan, Miranda, and their dad made a pie. They want equal shares. How should they cut the pie?

Name _____

Parts of Groups

Color to show each fraction.

1.

$\frac{1}{2}$

2.

$\frac{1}{4}$

3.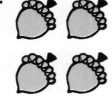

$\frac{1}{2}$

4.

$\frac{1}{3}$

▶ **Problem Solving**   **Animals circled may vary.**

Circle the animals to show the fraction.

5. There are 6 puppies.
   How many puppies
   are $\frac{1}{3}$ of the group?

6. There are 4 kittens.
   How many kittens
   are $\frac{1}{2}$ of the group?

Sort and Classify

This table shows one way
these toys can be sorted.

Cars and Trucks	
cars	|||||	
trucks	|||||

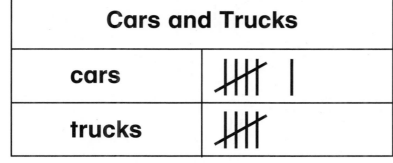

Each | stands for 1 toy.

||||| stands for 5 toys.

Sort the toys another way.
Make a table. **Answers will vary.**

▶ Problem Solving

Where is this fish shown
on the table? Circle the row.

Kinds of Fish	

Certain or Impossible

Look at the pocket.

Circle the pictures
that show what
can come out of it.

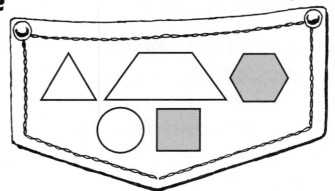

1.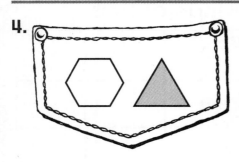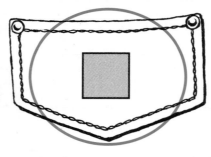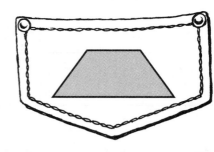

2.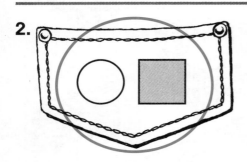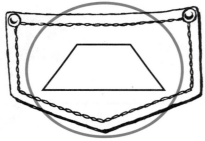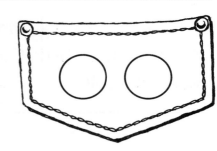

3.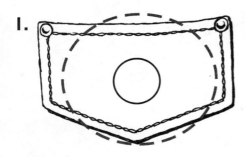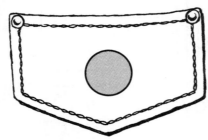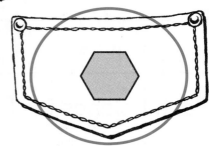

4.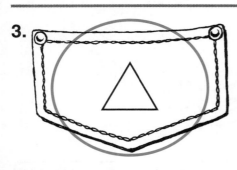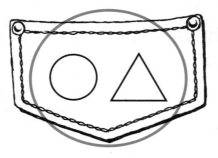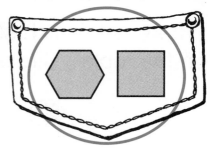

Most Likely

You will need: I bag, 6 red cubes,
5 blue cubes, and I yellow cube

Put the cubes into the bag.
Take out one cube.

Make a tally mark on the table
to show which color you got.

Put the cube back into the bag.
Shake. Make a prediction.
If you do this 9 more times,
which color do you think you will
get most often? Circle that color.

Color the cubes
red, blue, and yellow.

 red   **blue** yellow

Do this 9 more times.

Make a tally mark each time.
Count the tally marks for
each color.

Write the totals.

Answers will vary.

	Tally Marks	Total
red		
blue		
yellow		

▶ **Problem Solving**

Can you take a △ out of the 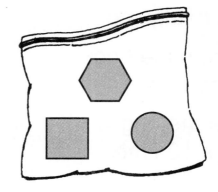 ?

Circle **Yes** or **No.**

Yes   No

Tallying Events

You will need: I bag, 5 red cubes, 3 blue cubes

Color the cubes. Predict which color cube you will get more often. Circle that cube.

Answers will vary.

Try it. Put all the cubes into the bag. Color the cubes in the table. Take out a cube. Make a tally mark.

Put the cube back into the bag.

Shake the bag. Do this I0 times.

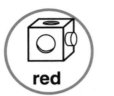
red

blue

	Tally Marks	Total
red		
blue		

Try again with these cubes.

You will need: I bag, 2 red cubes, 7 blue cubes

Color the cubes. Predict which color cube you will get more often. Circle that cube.

Answers will vary.

Try it. Color the cubes in the table. Put all the cubes into the bag. Take out a cube. Make a tally mark.

Put the cube back into the bag.

Shake the bag. Do this I0 times.

red

blue

	Tally Marks	Total
red		
blue		

Harcourt Brace School Publishers

Picture Graphs

bear	🧸	🧸	🧸	🧸	🧸
car	🚗	🚗			
doll	🎎	🎎	🎎	🎎	

Count the toys. Draw ◯ to fill in the graph.

	Our Favorite Toys				
bear	◯	◯	◯	◯	◯
car	◯	◯			
doll	◯	◯	◯	◯	

Use the graph to answer the questions.

1. How many cars are there? ___2___

2. How many more bears than cars are there? ___3___
_____

3. Are there more dolls or cars? - - - - - - - - - - - - - - - - - -
___dolls___

▶ **Problem Solving**

Look at the graph.

4. Write a number sentence
that tells how many bears
and dolls there are.

__5__ (+) __4__ = __9__

__9__ bears and dolls

Horizontal Bar Graphs

A group of children voted for their
favorite sports. Write how many tally marks.

Favorite Sport		Total
soccer	⊮⊮ ‖	7
softball	⊮⊮	5
football	‖‖	3

Color the graph to match the tally marks.

Favorite Sport							
soccer							
softball							
football							
	0	1	2	3	4	5	6	7

Use the graph to answer the questions.

1. How many children voted for soccer?   __7__

2. How many more voted for soccer than for football?   __4__

3. How many more voted for soccer than for softball?   __2__

▶ **Problem Solving**

Look at the graph.

4. Write a number sentence that
tells how many more children
like soccer than like football.

__7__ ⊖ __3__ = __4__

__4__ more children

Vertical Bar Graphs

Write how many tally marks.
Color the graph to match the
tally marks.

Where We Went for Vacation		Total
beach	|||| ||||	9
city	||||	5
farm	||||	4

Where We Went for Vacation

	beach	city	farm
10			
9	▓		
8	▓		
7	▓		
6	▓		
5	▓	▓	
4	▓	▓	▓
3	▓	▓	▓
2	▓	▓	▓
1	▓	▓	▓
0			
	beach	city	farm

Use the graph to answer
the questions.

1. How many children went to the beach?   9

2. How many children went to the farm?   4

3. How many more children went to the beach
   than to the farm?   5

▶ **Problem Solving**

4. Circle the question you can answer by reading the graph.

   How many children
   went to the mountains?

   (How many children
   went to the city?)

Name ____________________

LESSON 24.4

Problem Solving • Make a Graph

Ask 10 classmates to choose their favorite color.

1. Make a tally mark for each choice. Then write how many.

2. Fill in the graph. First write the title. Then color the graph to match the tally marks.

Our Favorite Color		Total
red		
blue		
green		
yellow		

Our Favorite Color

8				
7				
6				
5				
4				
3				
2				
1				
0	red	blue	green	yellow

Use the graph to answer the questions. **Answers will vary.**

3. Which color do the most children like best? ____________

4. How many children like yellow the best? ______

5. Write a question someone can answer by reading this graph.

____

Answers will vary.

Doubles Plus One

1. Circle the **doubles** fact.

$(4 + 4 = 8)$ $4 + 3 = 7$ $4 + 5 = 9$

2. Circle the **doubles plus one** fact.

$4 + 4 = 8$ $4 + 3 = 7$ $(4 + 5 = 9)$

Write the sums.

3. $3 + 3 = 6$,   so $3 + 4 = \underline{7}$.

4. $8 + 8 = 16$,   so $8 + 9 = \underline{17}$.

5. $6 + 6 = 12$,   so $6 + 7 = \underline{13}$.

6. $2 + 2 = 4$,   so $2 + 3 = \underline{5}$.

7. $4 + 4 = 8$,   so $4 + 5 = \underline{9}$.

Write the sums.

8.
$$\begin{array}{r} 7 \\ +7 \\ \hline 14 \end{array} \qquad \begin{array}{r} 8 \\ +9 \\ \hline 17 \end{array} \qquad \begin{array}{r} 5 \\ +5 \\ \hline 10 \end{array} \qquad \begin{array}{r} 5 \\ +6 \\ \hline 11 \end{array} \qquad \begin{array}{r} 4 \\ +4 \\ \hline 8 \end{array} \qquad \begin{array}{r} 1 \\ +1 \\ \hline 2 \end{array}$$

▶ **Problem Solving**

Use counters to solve. Draw them.

Check children's work.

9. Bill has 5 marbles. He finds 6 more marbles. How many marbles does he have in all?

$\underline{11}$ marbles

Doubles Minus One

1. Circle the **doubles** fact.

$3 + 4 = 7$ $\boxed{3 + 3 = 6}$ $3 + 2 = 5$

2. Circle the **doubles minus one** fact.

$3 + 4 = 7$ $3 + 3 = 6$ $\boxed{3 + 2 = 5}$

Write the sums.

3. $4 + 4 = \underline{8}$ | $6 + 6 = \underline{12}$ | $3 + 3 = \underline{6}$

$4 + 3 = \underline{7}$ | $6 + 5 = \underline{11}$ | $3 + 2 = \underline{5}$

---

4. $5 + 5 = \underline{10}$ | $9 + 9 = \underline{18}$ | $7 + 7 = \underline{14}$

$5 + 4 = \underline{9}$ | $9 + 8 = \underline{17}$ | $7 + 6 = \underline{13}$

5.

9	2	7	8	6	5
$+8$	$+3$	$+8$	$+9$	$+7$	$+6$
17	5	15	17	13	11

6.

7	5	4	4	8	3
$+6$	$+4$	$+3$	$+5$	$+7$	$+2$
13	9	7	9	15	5

▶ **Problem Solving**

7. Sue has 4 pennies. She finds 5 more. How many does she have in all?

_____ 9 _____ pennies

Check children's work.

Doubles Patterns

Write the sums.

doubles	doubles − 1	doubles + 1
1. 4 + 4 = __8__	4 + 3 = __7__	4 + 5 = __9__
2. 7 + 7 = __14__	7 + 6 = __13__	7 + 8 = __15__
3. 9 + 9 = __18__	9 + 8 = __17__	9 + 10 = __19__
4. 6 + 6 = __12__	6 + 5 = __11__	6 + 7 = __13__
5. 3 + 3 = __6__	3 + 2 = __5__	3 + 4 = __7__
6. 5 + 5 = __10__	5 + 4 = __9__	5 + 6 = __11__
7. 8 + 8 = __16__	8 + 7 = __15__	8 + 9 = __17__

▶ **Problem Solving**

Look at the coins. Write the number sentence.

8. + =

__5__ ¢ + __5__ ¢ = __10__ ¢

9. +  =

__5__ ¢ + __6__ ¢ = __11__ ¢

Doubles Fact Families

Add or subtract.

1.
$$\begin{array}{r} 5 \\ +\,5 \\ \hline 10 \end{array}$$
$$\begin{array}{r} 10 \\ -\,5 \\ \hline 5 \end{array}$$
$$\begin{array}{r} 4 \\ +\,4 \\ \hline 8 \end{array}$$
$$\begin{array}{r} 8 \\ -\,4 \\ \hline 4 \end{array}$$
$$\begin{array}{r} 7 \\ +\,7 \\ \hline 14 \end{array}$$
$$\begin{array}{r} 14 \\ -\,7 \\ \hline 7 \end{array}$$

2.
$$\begin{array}{r} 6 \\ +\,6 \\ \hline 12 \end{array}$$
$$\begin{array}{r} 12 \\ -\,6 \\ \hline 6 \end{array}$$
$$\begin{array}{r} 1 \\ +\,1 \\ \hline 2 \end{array}$$
$$\begin{array}{r} 2 \\ -\,1 \\ \hline 1 \end{array}$$
$$\begin{array}{r} 3 \\ +\,3 \\ \hline 6 \end{array}$$
$$\begin{array}{r} 6 \\ -\,3 \\ \hline 3 \end{array}$$

3.
$$\begin{array}{r} 2 \\ +\,2 \\ \hline 4 \end{array}$$
$$\begin{array}{r} 4 \\ -\,2 \\ \hline 2 \end{array}$$
$$\begin{array}{r} 9 \\ +\,9 \\ \hline 18 \end{array}$$
$$\begin{array}{r} 18 \\ -\,9 \\ \hline 9 \end{array}$$
$$\begin{array}{r} 8 \\ +\,8 \\ \hline 16 \end{array}$$
$$\begin{array}{r} 16 \\ -\,8 \\ \hline 8 \end{array}$$

4.
$$\begin{array}{r} 8 \\ -\,4 \\ \hline 4 \end{array}$$
$$\begin{array}{r} 4 \\ +\,4 \\ \hline 8 \end{array}$$
$$\begin{array}{r} 4 \\ -\,2 \\ \hline 2 \end{array}$$
$$\begin{array}{r} 2 \\ +\,2 \\ \hline 4 \end{array}$$
$$\begin{array}{r} 10 \\ -\,5 \\ \hline 5 \end{array}$$
$$\begin{array}{r} 5 \\ +\,5 \\ \hline 10 \end{array}$$

▶ **Problem Solving**

5. I had 18 pennies.
I lost 9 of them.
How many pennies
do I have left?

____9____ pennies

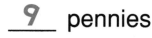

Check children's work.

Problem Solving • Make a Model

Use counters to solve. Draw them.

Check children's work.

1. Robert has 7 pencils.
   Sara has 1 fewer than Robert.
   How many pencils
   do they have in all?

   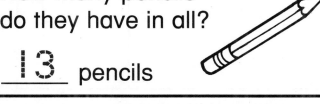

   __13__ pencils

2. Lisa has 4 dolls. Kara has
   two times as many.
   How many dolls do
   they have in all?

   __12__ dolls

3. John had 14 boats. He gave
   some of them away. He has
   7 left. How many did
   he give away?

   __7__ boats

▶ **Problem Solving**

4. I have 8 toy cars.
   My friend has the same
   number. How many cars
   do we have in all?

   Check children's work.

   __16__ cars

Make a 10

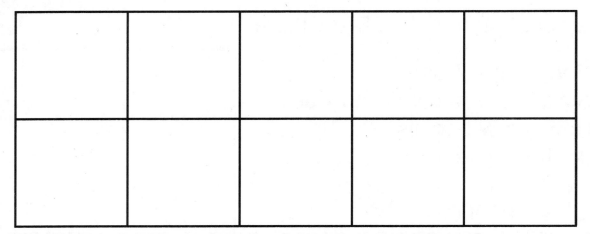

Use counters and the 10-frame. Start with the
greater number. Make a 10. Then add.

1.

3	8	5	3	9	6
+9	+5	+9	+8	+7	+8
12	13	14	11	16	14

2.

8	9	7	4	3	6
+7	+3	+9	+8	+9	+7
15	12	16	12	12	13

3.

8	7	3	3	9	7
+6	+4	+8	+9	+6	+5
14	11	11	12	15	12

Check children's work.

▶ **Problem Solving**

4. Mack had 9 pencils.
His dad gave him 2 more.
How many pencils does
he have in all? ___11___ pencils

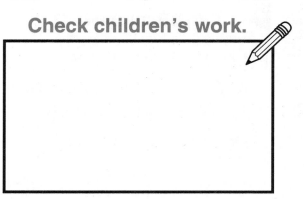

Harcourt Brace School Publishers

Adding Three Numbers

Circle names for 10 or doubles. Then add.

1.
```
   1          ③          ⑥          ⑦          ②
  ⑧          ③          5          ⑦          ⑧
 +⑧         + 6         +④         + 2         + 5
 ――         ――         ――         ――         ――
 17          12          15          16          15
```

2.
```
  ③          1          ⑧          ⑦          ⑤
  ⑦          ⑦          ②          4          ⑤
 + 2         +⑦         + 1         +③         + 1
 ――         ――         ――         ――         ――
  12          15          11          14          11
```

3.
```
  ⑨          ②          ⑨          ⑥          ④
  2          ⑧          3          ⑥          3
 +①         + 6         +①         + 3         +④
 ――         ――         ――         ――         ――
  12          16          13          15          11
```

Check children's work.

▶ **Problem Solving**

4. Jan has 8 yellow leaves,
   2 red leaves, and 6 brown
   leaves. How many leaves
   does she have in all?

   __16__ leaves

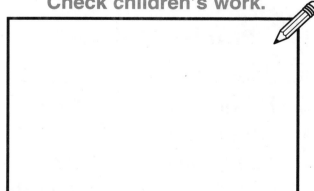

Sums and Differences to 14

Use Workmat 2 and counters. Add or subtract.

1.

```
   9        10          8        11          5        10
 + 1       - 1        + 3       - 3        + 5       - 5
 ----      ----       ----      ----       ----      ----
  10         9          11        8          10        5
```

2.

```
   8        13          7        14          6        12
 + 5       - 5        + 7       - 7        + 6       - 6
 ----      ----       ----      ----       ----      ----
  13         8          14        7          12        6
```

3.

```
   9        14          4        10          9        12
 + 5       - 5        + 6       - 6        + 3       - 3
 ----      ----       ----      ----       ----      ----
  14         9          10        4          12        9
```

4.

```
   9        11          8        14          5         9
 + 2       - 2        + 6       - 6        + 4       - 4
 ----      ----       ----      ----       ----      ----
  11         9          14        8          9         5
```

▶ **Problem Solving**

5. Joey has 14 stickers.
He gives 6 to Randy.
How many stickers
does Joey have left?

__8__ stickers

Check children's work.

Sums and Differences to 18

Write the sum and difference for each pair.

1.

$$\begin{array}{r} 9 \\ +9 \\ \hline 18 \end{array} \qquad \begin{array}{r} 18 \\ -9 \\ \hline 9 \end{array}$$

$$\begin{array}{r} 8 \\ +6 \\ \hline 14 \end{array} \qquad \begin{array}{r} 14 \\ -6 \\ \hline 8 \end{array}$$

$$\begin{array}{r} 6 \\ +6 \\ \hline 12 \end{array} \qquad \begin{array}{r} 12 \\ -6 \\ \hline 6 \end{array}$$

2.

$$\begin{array}{r} 8 \\ +3 \\ \hline 11 \end{array} \qquad \begin{array}{r} 11 \\ -3 \\ \hline 8 \end{array}$$

$$\begin{array}{r} 9 \\ +4 \\ \hline 13 \end{array} \qquad \begin{array}{r} 13 \\ -4 \\ \hline 9 \end{array}$$

$$\begin{array}{r} 7 \\ +6 \\ \hline 13 \end{array} \qquad \begin{array}{r} 13 \\ -6 \\ \hline 7 \end{array}$$

3.

$$\begin{array}{r} 9 \\ +7 \\ \hline 16 \end{array} \qquad \begin{array}{r} 16 \\ -7 \\ \hline 9 \end{array}$$

$$\begin{array}{r} 6 \\ +5 \\ \hline 11 \end{array} \qquad \begin{array}{r} 11 \\ -5 \\ \hline 6 \end{array}$$

$$\begin{array}{r} 6 \\ +8 \\ \hline 14 \end{array} \qquad \begin{array}{r} 14 \\ -8 \\ \hline 6 \end{array}$$

4.

$$\begin{array}{r} 6 \\ +9 \\ \hline 15 \end{array} \qquad \begin{array}{r} 15 \\ -9 \\ \hline 6 \end{array}$$

$$\begin{array}{r} 9 \\ +3 \\ \hline 12 \end{array} \qquad \begin{array}{r} 12 \\ -3 \\ \hline 9 \end{array}$$

$$\begin{array}{r} 6 \\ +7 \\ \hline 13 \end{array} \qquad \begin{array}{r} 13 \\ -7 \\ \hline 6 \end{array}$$

▶ **Problem Solving**

Check children's work.

5. I found 17 shells. I gave away 9. How many shells do I have left?

___8___ shells

Counting Equal Groups

Check children's work.

Use counters. Draw them. Write how many in all.

1. Make 2 groups. Put 3 counters in each group.

How many in all? __6__

2. Make 3 groups. Put 2 counters in each group.

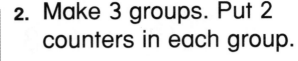

How many in all? __6__

3. Make 2 groups. Put 4 counters in each group.

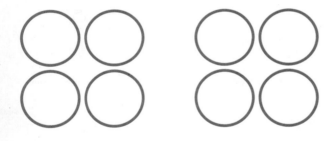

How many in all? __8__

4. Make 4 groups. Put 3 counters in each group.

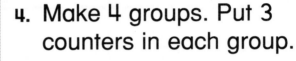
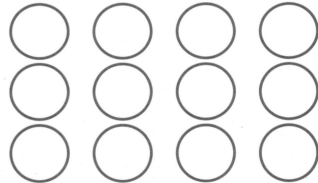

How many in all? __12__

▶ **Problem Solving**

5. Jessica has 3 dogs. She gave them 2 bones each. How many bones did Jessica need?

__6__ bones

Check children's work.

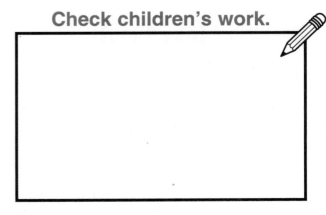

How Many in Each Group?

Use counters. Draw them. Write how many in each group.

1. Use 10 counters. Make 5 equal groups.

  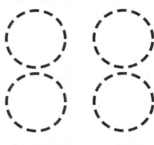

Check children's work.

How many in
each group? __2__

_____

2. Use 12 counters. Make 2 equal groups.

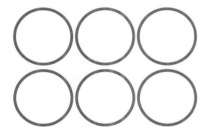 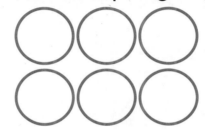

How many in
each group? __6__

▶ **Problem Solving**

Check children's work.

3. Dylan gave 4 pictures to his
brother and sister. He gave each
the same number. How many
pictures did he give to each?

sister __2__ brother __2__

_____

4. Jenny's grandmother gave 12
cookies to 3 children. She
gave each child the same
number of cookies. How many
cookies did each child get?

__4__ cookies

How Many Groups?

Check children's work.

Use counters. Draw them. Write how many groups.

1. Use 12 counters.
Put 3 in each group.

How many groups? __4__

2. Use 8 counters.
Put 2 in each group.

How many groups? __4__

▶ **Problem Solving**

Check children's work.

3. The teacher had 6 apples.
She put 2 apples on each
plate. How many plates did
she use? __3__ plates

4. Jason had 10 sugar cubes.
He gave 5 to each horse.
How many horses did
he feed?

__2__ horses

Problem Solving • Draw a Picture

Draw a picture to solve each problem. **Check children's work.**

1. There are 2 children.
   Each child has 3 balloons.
   How many balloons are there?

   _____**6**_____ balloons

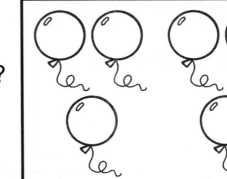

2. We have 3 pennies.
   We get 2 more.
   How many pennies
   in all?

   _____**5**_____ pennies

3. There are 4 baskets.
   Each basket has 3 eggs in it.
   How many eggs are there?

   _____**12**_____ eggs

4. There are 6 crayons.
   Each child gets 2 crayons.
   How many children
   get crayons?

   _____**3**_____ children

Adding and Subtracting Tens

▶ **Vocabulary**

Circle the Workmat that shows 4 **tens.**

tens	ones
	▱ ▱
	▱ ▱

tens	ones
▤▤▤▤	

Add or subtract. Use tens and Workmat 3.

1.

$$\begin{array}{r} 40 \\ + 20 \\ \hline 60 \end{array} \qquad \begin{array}{r} 70 \\ + 10 \\ \hline 80 \end{array} \qquad \begin{array}{r} 50 \\ + 20 \\ \hline 70 \end{array} \qquad \begin{array}{r} 30 \\ + 20 \\ \hline 50 \end{array} \qquad \begin{array}{r} 30 \\ + 10 \\ \hline 40 \end{array}$$

2.

$$\begin{array}{r} 40 \\ - 30 \\ \hline 10 \end{array} \qquad \begin{array}{r} 90 \\ - 30 \\ \hline 60 \end{array} \qquad \begin{array}{r} 70 \\ - 30 \\ \hline 40 \end{array} \qquad \begin{array}{r} 80 \\ - 20 \\ \hline 60 \end{array} \qquad \begin{array}{r} 60 \\ - 40 \\ \hline 20 \end{array}$$

Check children's work.

▶ **Problem Solving**

3. Ann had 40¢ to buy a pencil. The pencil cost 30¢. How much money does Ann have left?

___10___ ¢

Harcourt Brace School Publishers

Adding Tens and Ones

Add.

1.

tens	ones
3	2
+ 4	1
7	3

tens	ones
5	7
+ 3	1
8	8

tens	ones
4	2
+ 2	1
6	3

tens	ones
6	4
+ 3	5
9	9

2.

tens	ones
1	6
+ 2	3
3	9

tens	ones
7	4
+ 1	2
8	6

tens	ones
2	3
+ 3	5
5	8

tens	ones
1	2
+ 6	5
7	7

3.

tens	ones
3	1
+ 3	4
6	5

tens	ones
3	5
+ 4	3
7	8

tens	ones
8	5
+ 1	3
9	8

tens	ones
1	7
+ 3	2
4	9

▶ **Problem Solving**

Check children's work.

4. Elaine wants to buy a book and a pencil. How much money does she need?

 75¢

 12¢

___87___ ¢

Name_____

Subtracting Tens and Ones

Subtract.

1.

tens	ones
4	2
− 3	2
1	0

tens	ones
5	2
− 3	1
2	1

tens	ones
7	6
− 5	4
2	2

tens	ones
6	2
− 2	0
4	2

2.

tens	ones
5	4
− 2	1
3	3

tens	ones
9	6
− 1	1
8	5

tens	ones
7	3
− 4	3
3	0

tens	ones
2	2
− 1	1
1	1

3.

tens	ones
5	7
− 2	5
3	2

tens	ones
8	6
− 1	3
7	3

tens	ones
7	2
− 3	2
4	0

tens	ones
4	2
− 3	2
1	0

Check children's work.

▶ **Problem Solving**

Circle the answer.

4. Jessica had 75 buttons. She gave 12 buttons to Tyler. How many buttons does she have left?

54   36   (63)

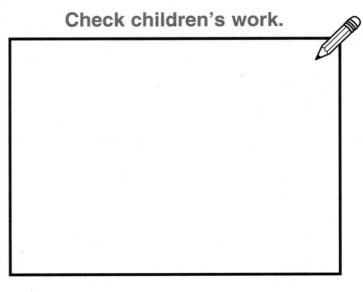

Harcourt Brace School Publishers

Reasonable Answer

Circle the answer that makes sense.

1. Chris had 17 books.
He gave 12 to the library.
How many books does
Chris have now?

   55 books

   (5 books)

   555 books

2. There were 35 children at
the zoo. 12 children went
home. How many children
are left?

   650 children

   76 children

   (23 children)

3. Steve had 13 stickers.
Then he bought 20 more.
How many stickers does
he have now?

   5 stickers

   133 stickers

   (33 stickers)

4. Rose saw 6 squirrels on the
fence. Then she saw 12
more in the grass. How many
squirrels were there in all?

   (18 squirrels)

   6 squirrels

   118 squirrels

▶ **Problem Solving**

Circle the answer.

5. Juan has 35¢.
He wants to buy a toy
car that costs 22¢.
How much money will
he have left?

  35¢   22¢   (13¢)

6. Sandy had 45 pennies.
Her mother gave her
23 more. How many
pennies does Sandy
have now?

  21¢   45¢   (68¢)